Three Dimensional Trimming and Machining

The Five Axis CNC Router

FIRST EDITION

Thermwood Corporation
Old Buffaloville Road
Dale, Indiana 47523

Printed in the United States of America
First Edition Printed 1999

ISBN 0-9665693-1-8

Table of Contents

Introduction

Machining and trimming three-dimensional parts has become a major part of several industries. The thermoforming industry makes extensive use of CNC routers to trim three-dimensional formed parts. As this book is being written, pattern making is being transformed with the use of CNC routers. Rotational molding, blow molding and even injection molding are all starting to make use of this technology. In the wood industry, in addition to patterns, formed plywood parts, typically seating is being sized using CNC routers.

I am president of a company that builds CNC routers. As such, I am obviously quite biased toward the product that my company builds. At the same time I recognize that the major deterrent to increased use of these machines is a general lack of understanding of the basic technology and operating techniques. This shortcoming in the market makes the learning curve for new users much longer and more difficult than it needs to be. It also results in experienced companies only getting a portion of the available benefits from their CNC router investment.

The basic facts, technologies and techniques are universal. They do not apply to just a single brand machine. This means that it is possible, at least in theory, to discuss the fundamental technology without it becoming a sales pitch. I will make a sincere attempt to thoroughly discuss the technology without trying to promote the products that my company makes.

At the same time, I am familiar with the intimate details of how we address and solve some of the technical challenges of building this type of machine. I am not as aware of all the techniques that my competitors use to address these same problems. Therefore, I will discuss the methods we employ to solve certain technical issues. In many cases there are other valid approaches to the same problem. If I know of these, I will try and explain them, however, if you are dealing with a machine supplier that does not use the same techniques that we do, it will be necessary for you to ask them specifically how they address each of these areas.

The first automatic routers were numerically controlled or NC. This means that they were controlled with numbers. These first appeared in the late fifties and early sixties. Their development as NC routers seems to have come from government sponsored R&D programs. These early machines generally had two axis of motion. The third axis, (the "Z" axis) raised and lowered the router head and was on an air slide with an adjustable stop at the bottom. The machines were actually controlled by a punched tape, which was essentially a paper strip about three quarters of an inch wide with holes punched to correspond to the control numbers. These controls did not have a computer, but were simply an electronic control that responded to the number punched into the tape. You might think of it as a fancy player piano.

In these early days, the number of feet of tape required to run the program determined program size. Control capacity was listed as the maximum number of feet of program tape that could be run.

By the early seventies, companies in the aerospace industry were experimenting with mini-computers to control machine tools. Numerically controlled (NC) became computer numerically controlled (CNC). CNC offered the potential for easier operation and more capability, however, the cost of the control was very high.

In the mid-seventies, Thermwood Corporation developed the first CNC control based on the then new, potentially low cost microprocessor. Although this control was crude by today's standards, it did offer some major technical advances over conventional NC such as automatic acceleration and deceleration. We will talk about some of these features later in the book.

At the time, Thermwood was in the plastic molding business. They were thermoforming parts for the furniture industry. The initial machines that were built were not intended for sale but were to be used in making plastic parts. By the late seventies, Thermwood realized that to trim three-dimensional formed parts they would need a five-axis machine. They built the first five-axis CNC router patterned after multi-million dollar five-axis machines in the aerospace industry.

At about this same time, Thermwood saw a strong downturn in their plastic molding business as the furniture industry was moving away from plastic and back to wood as their primary material. They decided to sell CNC routers to other plastic molders to supplement their shrinking revenue from plastic parts.

Eventually, they sold off their plastic molding operations to focus exclusively on CNC routers. Commercial controls that could address all five axes became available and other companies also began to offer five-axis CNC routers.

Today there are several companies offering five-axis CNC routers. A couple of these companies have been in the business for many years while others are just joining the competition with new offerings. With the proliferation of suppliers, some experienced and some not, the need for a comprehensive review of existing technology, methods, capabilities and limitations has become even greater. With this book I am attempting to address this need.

The CNC router offers many benefits to the plastic molder. In general, without automatic machines, the parts will need to be trimmed by hand. This results in a huge number of ongoing problems. The first problem is labor. With a growing economy and a labor force that is growing at a slower and slower pace, labor is becoming more and more expensive. Even worse, sometimes it is not possible to even hire people to do this type of work. When you rely on manual labor for this job and you can't hire manual labor you are in trouble.

Using manual fixtures to guide a hand router, knives or other trimming devices is physically demanding. Because the operator is in close contact with the trim device, accidents are common. Most of the time these accidents are minor in nature consisting of cuts, scrapes and gouges. Almost always they result in a lost time accident which our friends at OHSA watch very carefully. Sometimes the accidents are even more serious resulting in serious, permanent injury.

Even without injury, other problems are looming on the horizon. The repetitive nature of the trimming task can result in carpal tunnel syndrome or other repetitive motion injuries. Our legislatures have been looking at these problems and several bills have been introduced to try and limit the amount of time each day that a person can be required to do this type of work. I am not aware of any of these bills having been voted into law, but the handwriting is on the wall. When this type of legislation finally gets passed, it will make an already difficult situation even worse.

Even without legislation, I am aware of companies that spend literally millions of dollars a year treating repetitive motion problems. Even without the economic considerations, there is the human consideration. Repetitive hand trimming will eventually hurt most people. If a viable alternative exists there is no good reason to continue harming our employees.

Today, however, the marketplace may actually dictate the methods used. Parts trimmed on a CNC router are of substantially higher quality and consistency than parts trimmed by hand. Companies with the proper equipment are at a competitive advantage. I am aware of several instances where the end customer requires, in their purchase orders, that the parts they purchase be trimmed on a CNC router.

With all these forces at work, we are rapidly moving toward the time where automatic trimming with a CNC router is the normal, accepted manufacturing process. Guide fixtures and hand trimming will become a smaller and smaller part of our industry and companies without CNC routers will not be competitive.

At the same time, CNC router technology has been advancing. Machine prices have been decreasing even in the face of inflation. Control systems have become substantially more capable. Programming three-dimensional parts, which has been a very difficult, tedious process in the past, has now become both fast and easy. CAD/CAM systems that were literally priced in the millions of dollars can now be purchased for a few thousand dollars, and they perform better than the multi-million dollar predecessors.

Just as the need for CNC routers is becoming stronger and stronger, the product is becoming advanced enough to handle most requirements. They are becoming low enough in cost that most companies can afford them. They are becoming simple enough to use that normal manufacturing companies can deal with and master their operation. The CNC router is becoming a common tool.

Like most tools, however, the final result depends on both the tool and the skill of the person holding the tool. The skill in using a CNC router varies widely. There is one major difference between the CNC router and other tools. With most other tools there are numerous books and self-help guides showing you exactly how to get the most from their use. The techniques and methods can be learned from published literature. At least today, no such help exists for using CNC routers.

That is where this book comes in. I hope that this book can fill the void. I am intending it to be a fundamental guide. Today, companies that want to run CNC routers must learn from their machine supplier, from any current machine user that is willing to share their experience and from trial and error. I have seen numerous cases where companies use cumbersome procedures where a simple straightforward method exists. They simply don't know the technique.

Even more serious, I have seen instances where companies developed attitudes toward both the machine and the expected performance that were not accurate. They imagine how a particular task should be completed. Then they go through incredible efforts and frustration trying to implement their idea. They usually fail, blaming the machine, the tooling and their suppliers, when all the time it was their fundamental concept that was flawed.

In these instances, there is usually a method of achieving the result the company desires, it is simply not the method they imagine. I have seen numerous instances where a company refuses to modify their methods to achieve their objective. Instead, they stubbornly insist that not only do they want a specific result, but they want to achieve that result using a specific approach. When the approach they insist on will simply not achieve the result they want, the installation will fail. This is frustrating as a supplier to the industry and the failures unnecessarily damage the image of the technology.

I hope that this book will help manage expectations so that fewer of these problems will develop. If we can help prevent inexperienced companies from developing inaccurate expectations, this effort will be a success. By showing the multitude of techniques and the breath of the available technology, I hope to show companies that there is no reason to fixate on one approach. There are numerous ways of approaching a requirement. If one doesn't work, simply try another and another until a successful technique is found.

The good news is that a company with a clear idea of what they want to accomplish and a willingness to use whatever technique will work is virtually guaranteed to succeed.

To try and manage expectations and provide customers with an accurate image of what they can expect, we have operated a demonstration lab for many years. We try to fixture, program and run a product that the customer intends to produce prior to the purchase of the machine. The customer then has a realistic view of what to expect.

In actual practice, however, most customers send us their most difficult parts. They assume that if we can produce this part, we can likely produce anything they make. After years of this, the people in my company that are responsible for making this work have developed a tremendous level of skill and experience. It is this storehouse of knowledge that I have tapped for this book. In fact, the technology, techniques and knowledge that are contained in these pages did not exist within one person before it was collected and put down on paper.

I would like to take a moment and thank all those people that shared their knowledge and experience and helped make this work possible. I know that it does not contain all the answers. The underlying technology is changing rapidly. Had this book been written five years ago, fully half the technology covered would not have existed. Five years from now, I expect that a large portion of the available technology will be new.

Despite this, we must start somewhere. This is the start. I will try and thoroughly cover the core technology and concepts. I will try and do this in as non-technical and easy to understand terms as possible. If this technology is to become the useful tool that we will need in the future, it will be necessary for the end machine users to better understand the underlying technology. I hope this book proves to be an important step in that understanding.

Chapter 1

Machine Basics

Machine Basics

Why five axis? Why not four, why not six?

To understand this seemingly trivial question we must understand something about the geometry of trimming. First, what is an axis? From an engineering standpoint an axis is a freedom of movement. A drawer has a single axis of motion. The drawer moves linearly, in and out. This linear movement is a degree of freedom and in engineering terms is represents a linear axis.

A door hinge is also a single axis. In this case, however, the axis is a rotary axis instead of a linear axis. The door rotates around the hinge pin instead of sliding along a linear path. While there are other forms, these two (linear and rotary) are the two basic axis movements. They can be combined in a seemingly endless variety of combinations. From this point forward we will use these two axes, linear and rotary to describe various machine configurations.

The two different axes I have just described, the drawer and door, each have a degree of freedom, but the axes are not controlled. Once we attach some type of driving device to an axis it becomes a controlled axis. For example, we could attach an air cylinder to the drawer to open and close it. It is now a controlled axis. A stop at each end of the cylinder determines the open and closed position.

What if we wanted to stop the movement somewhere in-between open and closed? Or, what if we wanted to control the speed at which the drawer opened or closed? In either case, we must servo control the axis. What is servo controlled?

Servo control is a method by which you can define any point along an axis and then achieve that position. Normally this is done with a servo drive. Although there are other systems, today a servo drive is normally a special motor, (called a servomotor) with a position feedback system and an electronic drive. The position feedback system tells the control the rotation angle of the servomotor. The control then compares the actual position of the motor to the desired position. The difference between the desired position and the actual position is called an "error signal". This error signal is then amplified and directed to the motor to rotate in the direction necessary to reduce the error. As the motor nears the desired position, the error signal becomes smaller and causes the power fed to the motor to also reduce until the desired position is reached and the error signal becomes zero. At that point, no power is fed to the motor.

As soon as you try and rotate the motor from the desired position, an error signal develops which results in a motor torque opposing the movement. Thus, for all practical purposes, the motor is locked onto the position.

To create a programmed motion, the desired position is electronically "moved". The servo system then follows this moving point, resulting in the desired movement.

This description is extremely simplified and it misses many significant points. It does, however, offer the most basic view of how machine motion occurs. In the next chapter we will spend much more time discussing servo systems and the most common methods of achieving servo motion.

Now that we know what an axis is and more or less how it is moved, let us look at basic machine configurations. We will start with simple three axis machines.

Let us start with the basic machine configuration. CNC routers have been built in four different configurations over the years. Today, two of these dominate the three axis machine offerings. Each of these configurations can work equally well. Those companies that claim one is better than another are simply wrong. Each does have strong and weak points, however, if properly engineered each will produce high quality production.

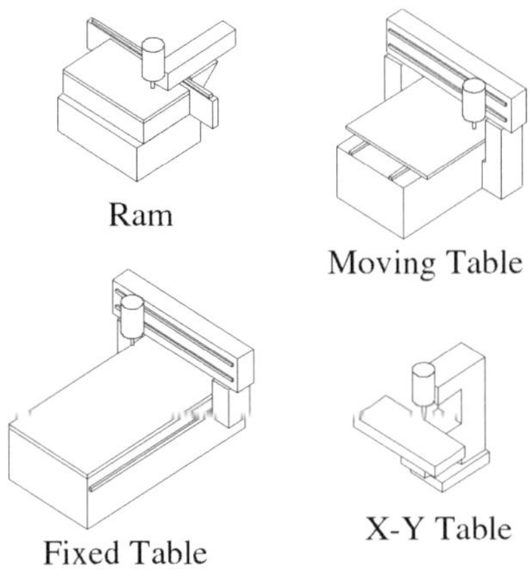

Common CNC Router Configurations

This can be very confusing if you listen to the desperate claims of some machine manufacturers that maintain that only one design will do your job. By coincidence this is the only design they build. These people will point out factors that must be taken into account when a configuration is chosen and show that it makes a particular configuration slow, inaccurate or unstable. The truth is that none of the configurations will work properly if they are designed incorrectly. The converse is also true, all will work very well if they are properly engineered. The basic configuration you select should be based on other factors than the belief that one is inherently better than another.

The first NC routers were a ram design. A tool carriage ran left and right on a set of rails located directly behind the worktable. The worktable was about two feet wide and eight to ten feet long. The operator stood with the ten-foot dimension to his left and right. A ram type structure was mounted to the carriage so that it could move from the back of the table, two-feet to the front of the table. The head could then cover the entire table, front to back and left to right.

This configuration has one major advantage. It is easy to load and unload. The table is suspended in front of the operator and all of the operating mechanism is located behind the table. Every part of the two-foot wide table can be easily reached.

Its major disadvantage is the table width. A part over two foot wide cannot be completely machined, although some companies machine wide parts by machining as much of the part as possible and then turning the part around to machine the other side. To try and overcome this shortcoming, this type of machine has been built with a wider table. Since the ram structure is suspended from only one side, developing a structure that remains rigid becomes quite difficult. In general, this is an area where you try and create a very rigid but lightweight structure.

Today, the ram style machine is commonly used for point-to-point machining centers, which are used in the wood and furniture industries but are seldom used for the somewhat more demanding CNC router applications.

Another machine style that is seldom used is the X-Y table. In this design, a table that moves in both right and left and front to back is mounted under a spindle that moves up and down. The first of these machines was actually a pin router with an X-Y table mounted to it.

It is quite easy to get a very rigid machine using this design, however, from a practical standpoint it is limited to rather small table sizes. The spindle must be attached to the machine base by an upright column. The distance from the column to the spindle defines the maximum table width and this distance cannot be too large without making the overall machine structure impractical.

The final two machine configurations, the fixed table/moving gantry and the moving table/fixed gantry designs are the most common in industry today. Properly applied, each design produces essentially the same results, but the design requirements for each are quite different.

The moving table machine is more common than the fixed table machine, not because it is inherently move stable as some manufacturers might have you believe. Instead it is more common because of a control system limitation.

To understand this, we must understand how modern machines achieve their accuracy. A ball screw and nut combination drive most machine axes. The nut is actually a bearing that turns on the screw with very little friction. The ball nut is mounted to the part of the machine that is to move and the screw is mounted to the base. As the screw turns, it moves the ball nut and the axis it is connected to.

This is simple enough until you think about accuracy. To accurately position the nut, the screw must be very accurate. If there is any error, (called pitch error) in the screw, the position that the control thinks the nut is at and its actual position will be different. So how do we get an accurate screw?

Screws are made using two different processes. The first is the rolled ball screw where the ball tracks are rolled into the screw shaft using a special machine. This produces a high quality screw but, because of the nature of the rolling process, one that is susceptible to pitch error. Even a high quality rolled ball screw can have a pitch error of .010 to .015 inch per foot of length. This means a ten-foot screw could be off by .150 inch or over an eighth of an inch. This is outside a practical tolerance for many applications.

The second type of screw is the ground lead screw. In a ground lead screw the track is rolled or machined into the rod and then precision ground to a high level of accuracy. A ground lead screw can be manufactured with a tolerance of only a few thousands of an inch per foot instead of ten to fifteen thousands per foot for a rolled lead screw.

There are two problems with this. First it is very very expensive. A ground lead screw can cost thousands and thousands of dollars. This might be alright if it were very accurate, but with only a few thousands of an inch error per foot the axis can still be off twenty to thirty thousands over ten feet. While this may be acceptable for most applications, it is still undesirable especially when you consider the money invested.

The answer to this problem has come from the control system manufacturers. The only problem with a rolled lead screw is that when the control thinks the axis is in one position, it is actually in a slightly different position depending on the pitch error of the screw. If the control knew the exact pitch error of the screw, it could allow for that error when positioning the axis achieving accuracy well beyond the accuracy of the screw.

The feature that performs this correction is called lead screw compensation. The control system maintains a lead screw compensation table that is actually a map of the differences between where the control thinks the axis is and where it actually is. This table is created after the machine is built, by moving through the entire range of motion and comparing the actual position of each axis with the theoretical position. The most accurate way to measure the actual position is by using a laser interferometer. This is a measurement device that can measure to an accuracy in the millionths of an inch over large distances. This process is so sensitive that factors such as temperature, humidity and barometric pressure must be taken into account because they slightly affect the speed of light which can affect measurements this small.

Lead screw compensation tables can also be created using a reference screw. This is either a very accurate screw or one that has been previously mapped using the laser. This process is normally done to the screw before it is assembled to the machine and has the disadvantage of only allowing for lead screw error. When the table is created on the finished machine it compensates for any machine factor that might affect accuracy. We will examine the entire question of machine accuracy and repeatability later in the book.

Each control system manufacturer offers a different number of compensation points per axis. Some offer a couple of hundred for the entire machine while others offer thousands per axis. The closer the compensation points are together the more accurately the screw is mapped, but the harder the control must work to keep tract of actual position.

The result of all of this is that using a rolled ball screw and lead screw compensation table we can achieve positioning accuracy that is much better than a much more expensive ground lead screw without compensation.

At times you may see someone using both a ground lead screw and a lead screw compensation table. Does this give even more accurate results as the manufacturer will likely claim? Actually no. Some older and some lower cost CNC controls limit the amount of compensation that can be entered into the compensation table. By limiting the size of the number, calculations can be performed using fewer bits reducing both cost and complexity. The practical result is that many times a rolled lead screw cannot be properly mapped within the limits of the control. The next best thing is to use a ground lead screw, however, it is not necessary to make the screw as accurate as possible. It is only necessary to make the screw accurate enough to fit within the mapping limits of the control. This can be much less expensive than a highly accurate ground lead screw. In general however, a ground lead screw is overkill for CNC router applications and results in a machine that is substantially more expensive without offering any measurable benefit for the extra cost.

There is a type of screw that is a combination of a rolled lead screw and a ground lead screw, called a polished lead screw. This type of screw is essentially a rolled lead screw from an accuracy standpoint. It is lightly ground or polished to give it a smoother track, which in theory offers a smoother motion than a simple rolled screw.

Lead screw selection is one of those engineering decisions that must be made. Do I use a more capable but more expensive control and lower cost rolled lead screws or do I use a low cost control but somewhat more expensive ground lead screws? The positioning results of both will be essentially the same.

Now that we understand lead screw compensation, let us turn back to machine configuration. A moving table machine has a single lead screw moving the head back and forth on the gantry and a single screw moving the table front to back. This is a straightforward arrangement.

A fixed table machine has the moving gantry mounted to a rail located on either side of the table. A lead screw along each rail normally drives this gantry. One lead screw moves the head back and forth on the gantry, but two lead screws are required to move the gantry itself. This is where it gets a little tricky.

Movement of the gantry is a single axis. This single axis, however, must drive two servomotors, not one. Each servomotor drives a lead screw. The chances of the two lead screws being exactly matched are virtually zero so a single compensation table won't work. Each screw must have its own independent compensation table. As you can see this can get quite complicated.

Control system manufacturers have created a package to address this requirement called "gantry software". This configuration is commonly used on multi-million dollar aerospace machine tools so the control manufacturers haven't been shy about its price. It is not uncommon for gantry software to double the cost of a control.

So, the fixed table machine requires one extra servomotor and drive, and requires gantry software making it more expensive. It is easy to understand why the moving table has become the most popular design.

There have been several attempts to work around these areas. The most direct approach has been to use a single drive motor running two screws through a jack shaft, belts or the like. The biggest concern with this arrangement is that it does not account for the differences in pitch error between the two screws. It is possible to select and match screws within a reasonable tolerance but this is a time consuming and difficult task. When this is done, the compensation table should be the average of the pitch error of the two screws making each side of the gantry off by half the difference.

It is also difficult to design a drive system to connect the two screws that provides a solid drive with no lash.

Another method that is starting to be used in fixed table designs is to cantilever the table, supporting it at each end and then connecting the legs of the gantry underneath the table. The gantry is then driven by a single screw in the middle, underneath the table eliminating the need for gantry software.

The only potential problem with this arrangement is that it is more difficult to achieve a rigid structure. There is a tendency for the gantry to skew when a heavy cut is taken at the edge of the table. The bearing and rail structure must be substantially stiffer for this arrangement to work as well as dual drives. Achieving this level of mechanical stiffness will add cost to the machine and at some point it makes more sense to go back and drive the gantry from both sides.

When talking to a new machine manufacturer, make certain that they are aware of, and address, the lead screw compensation issue. There are companies that simply ignore this and let the customer live with the results. In some cases they do this where the control is more than capable of providing the compensation function.

The moving table machine design and the moving gantry machine design achieve essentially the same results in a three-axis configuration if both are equally well engineered. The biggest practical difference between the two designs is that for equal table size, the moving table design requires about twice the floor space. The fixed table is made up of the table and enough area around the table for the gantry to ride. The moving table machine needs enough room for the table to move its entire length and requires at least twice the space. Also, because of the requirement for dual drives and gantry software, a fixed table design tends to be more expensive.

There are some other differences between the two designs. A fixed table machine can generally carry heavier parts than a moving table machine. Table movement must not only move the table but must also move the fixture and workpiece. When the fixture and workpiece weigh thousands of pounds this can become a problem. In these circumstances it is better to place the fixture and workpiece on a fixed table braced to the floor and move the gantry over the work. Large aerospace machines are always fixed table moving gantry designs. They have large tables and process heavy fixtures and parts.

As we move into five axis machines, both the moving table and moving gantry designs are commercially available. At least one manufacturer is currently offering a ram type machine with a dual rotary head although the vast majority of machines are of the fixed gantry, moving table design.

There are several major differences between the three-axis CNC routers and the five-axis machines. The biggest difference is that five-axis machines are generally designed to trim or machine three-dimensional parts where the three axis machines are intended to process primarily flat parts. This means that the five-axis machine must accommodate parts that are taller. The gantry, whether it is fixed or moving, must be taller and the vertical axis, generally called the "Z" axis, will have much more stroke than the three-axis machines.

The gantry of a five-axis machine generally allows for at least 24 inches of Z-axis stroke and sometimes Z-axis heights of 60 inches or more are used. Gantry height and Z-axis stroke are a two edged sword, however. The greater height does offer larger part processing capability but at a significant cost. As the gantry gets taller and the Z axis gets longer both accuracy and stability diminish rapidly.

It is extremely difficult to build a long rigid cantilevered Z axis that is rigid enough to accomplish rapid acceleration and deceleration without vibration. This axis vibration or shaking results in a poor edge cut quality. Cutting forces in most three-dimensional trimming applications are quite small so most of the shaking is induced through the drive. The situation can be significantly improved by reducing the acceleration and deceleration of the various axes that affect the gantry.

There are several factors that affect the cycle time or the time required to trim a part. Top speed is one of those factors, however, acceleration and deceleration will generally have more impact on cycle time than top feed speed. Acceleration is a measure of how fast the axis accelerates to top speed and deceleration is a measure of how fast an axis stops.

Reducing acceleration and deceleration to try and stabilize a tall gantry and long Z Axis will significantly slow the machine on all parts. For this reason alone it is best to try and keep the Z Axis and gantry height as low as possible, but there is another reason.

As the Z Axis becomes longer, it is increasingly more difficult to obtain consistent accuracy at the bottom of the stroke. Getting the long axis to travel in a perfect straight line is only the first problem. The Z Axis generally attaches to the cross axis on the gantry. It is normally carried by two linear rails that run along the gantry. If these two rails are not perfectly aligned, if they vary with respect to each other even slightly, the long Z Axis will magnify the error significantly.

If the machine is of a moving gantry design, the situation can be even worse. Not only does the Z Axis moving on the gantry tend to pitch with small alignment errors but the entire gantry as it moves along the base can also pitch and magnify errors. If you absolutely need the extra machine height, you can expect to either slow machine execution or live with a poorer quality cut. In either case, you can also expect lower accuracy than a similar machine with a shorter gantry. As a general rule, buy only the Z Axis and gantry height you need. Getting a taller gantry in case you might need it in the future can be very costly for current production.

Moving gantry five-axis machines are commonly used today. In general, this machine configuration is used when large table sizes are required or when heavy fixtures are needed. Standard machines are available with table sizes up to ten foot by twenty foot.

These machines will accelerate and decelerate slower than a fixed gantry, moving table machine. As the tall gantry is accelerated and decelerated, it tends to pitch forward and back. This tendency to pitch is resisted by the main linear bearings and rails mounted along the base. High acceleration and deceleration rates tend to put high loads on these bearings and rails. Even a slight flex of any of the components or mounting structure will result in a pitch of the gantry. Even a slight pitch can be magnified by the gantry height to an unacceptable level. The only sure way of preventing this is to slow the acceleration and deceleration rate to a point where no gantry pitching occurs.

In general, this means that a moving gantry machine will process parts slower than a moving table machine where the gantry is fixed and thus cannot pitch. Since most moving gantry machines are trimming parts that are either too large or too heavy for a moving table machine, this comparison isn't really that important.

Moving table machines tend to be less expensive than moving gantry machines, even with the same table size. The structure of a moving gantry machine is much more involved than a moving table machine. In addition, a moving gantry machine requires two servomotors to drive the gantry and it requires a CNC control with gantry software. All of these factors tend to make the fixed table, moving gantry machine more expensive.

In five-axis machines, in most cases if the part size and weight allows, a fixed gantry moving table machine will be the better selection. It will generally be a lower cost machine that trims parts faster.

Another major difference between three-axis CNC routers and five-axis CNC routers is the two rotary axes that comprise the head mechanism. Head configuration can get quite complex so we are going to look at it in some detail.

The head of a five-axis CNC router is generally made up of two rotary axes. There are actually two configurations in which a head can be arranged. According to EIA274D, there are three axes that can be configured in the head. EIA274D is an NC/CNC standard that almost all control and machine manufacturers follow. According to this standard there are three mutually perpendicular axes that make up the rotary head of a machine tool. These are called the "A", "B" and "C" axis.

The three linear axes, "X", Y" and "Z" combined with any two of the rotary axes will yield a machine that can place the head in any orientation with respect to the part. If all six axes are used, not only can the head approach the part from any orientation but it can also rotate about the center of the tool. On a CNC router this means that the router spindle can rotate about the axis that the tool is rotating about. Another way of imagining this is that if you lock the spindle you can rotate the tool using the head.

This additional degree of freedom is not useful in routing but is used extensively in industrial robots. On a robot, instead of a rotating cutter bit, there will be a gripper, spray gun or other tool. These tools must generally be rotated about the sixth axis to achieve the necessary orientation toward the part. For normal CNC router applications the sixth axis offers no benefit.

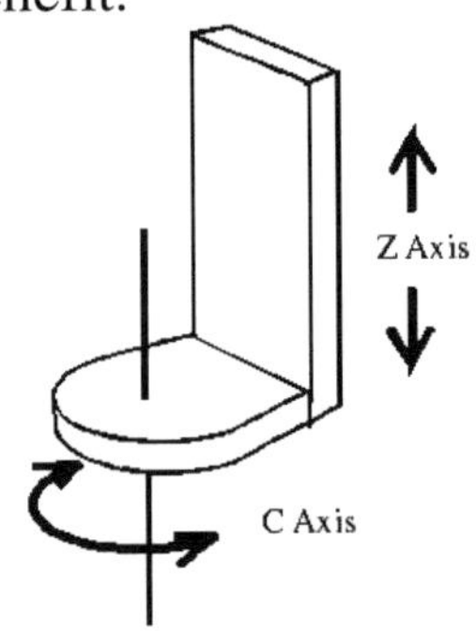

The "C" axis is defined as a rotary axis that is perpendicular to the "Z" axis with its axis of rotation parallel to the movement of the "Z" axis.

The “B” axis is the axis that has its axis of rotation along the “X” axis when in the reference or “home” position. The “C” axis is the axis that has its axis of rotation along the “Y” axis when in the “home” position.

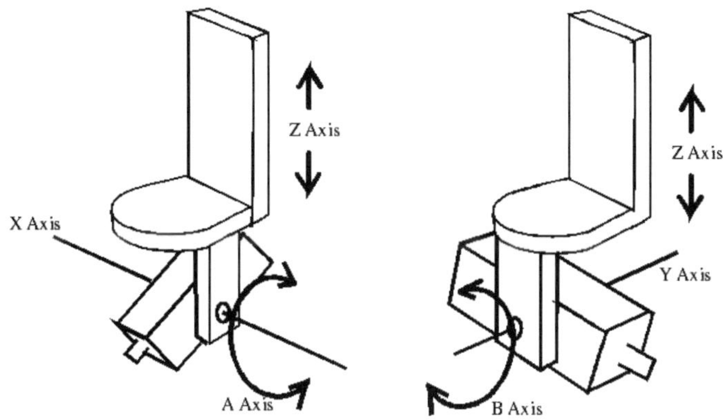

Most CNC routers use the “C” axis and either the “A” or “B” axis for their head configuration. It is also possible to build a head with an “A” and “B” axes but without a “C” axis. This configuration, however, is not nearly as easy to build as the traditional “C and A” or “C and B” heads. The “C and A” and “C and B”” heads are actually the same head. The only difference is the orientation of the head when in the home position.

We will talk about accuracy later in the book, however, setup and alignment of the two rotary axes of the head is a major consideration. Alignment of a five-axis head is a fairly complex effort that is difficult to comprehend. To obtain reasonable accuracy at the cutting tool, the head alignment must be extremely tight. The actual mechanical configuration of the head will greatly influence this effort.

Early five-axis CNC routers used two arms to mount the spindle cradle. Rotation of the spindle cradle is the A or B Axis sometimes called Axis 5. It was felt that two arms offered better stiffness and rigidity than a single arm arrangement. Actually it is quite a bit easier to obtain a rigid rotary axis using two arms.

This arrangement is not commonly used today because another problem developed. To properly align Axis 5 with the rest of the machine, the axis of rotation must be adjusted until it is exactly perpendicular to the C Axis, sometimes called Axis 4. When a two-arm mounting for Axis 5 is used, the shaft that rotates in one arm must be exactly in line with the shaft that rotates in the second arm. Getting two shafts precisely in line is an extremely difficult effort. We are trying to achieve alignments of a thousandth of an inch or so and there is no simple way to even measure the alignment. Even after the shafts have been aligned with each other, they must then be aligned with the remainder of the machine. In practice, achieving a reasonable level of alignment has proven to be quite difficult.

If the two mounting shafts of Axis 5 are not aligned with each other, both parallel and angular, the path traced by the rotation of the spindle is not a flat plane, perpendicular with Axis 4. Instead, it will trace an odd warped path that is all but impossible to predict. The amount of this warp is generally rather small, usually amounting to only a few thousandths of an inch. That few thousandths of an inch, however, may translate to twenty to thirty thousandths when projected out to the cutting tool tip. This can become a significant error, especially when trying to cut patterns using CAD/CAM generated data.

Because of this difficulty, machine manufacturers have generally turned to a single arm mounting for Axis 5. This arrangement is more difficult to engineer but it makes a more accurate alignment of Axis 5 somewhat easier to achieve. Alignment of a five-axis head is never easy, as you will see later in the book, but with a single arm mounting arrangement, it is normally possible to achieve a higher level of alignment than with a double arm mounting arrangement.

Now we are back to the original question asked at the beginning of this chapter. Exactly how many axes are required to do a particular job?

Many companies will be surprised to learn just how many jobs a three-axis machine can perform. A three-axis machine can move anywhere within the working envelope of the head but with only a single tool orientation. The orientation of the cutter head is fixed. Normally the head is mounted vertically with the tool pointing down. This tool can then be moved anywhere on the machine.

A lot of trimming and machining on plastic parts is done on the part line or mold line. This part or mold line is defined by the parting line of the mold that molded the part. This parting line must be essentially a plane for the mold to able to open and close. Also, unless you have a mechanically actuated mold, the parting line must be the outermost position on the part. Essentially this means that the part usually does not have undercuts. If it is a simple mold that does not have moving parts that allow for undercuts, the entire mold line can be reached using a three-axis machine.

This assumes that the trimming or machining required can be accomplished from above using a tool that is vertically mounted. In many cases, the tool must be oriented differently with respect to the part. If the number of different tool orientations required is small, it is possible to still use a three-axis machine and simply reorient the part instead of the tool. Once the number of orientations required to trim a part becomes large or the tool orientation required to trim a part must change continuously during the cut, additional axes must be added.

The next logical step is five axes. With five axes, it is possible to achieve any orientation of the tool. We can approach the part from any angle.

Essentially there are two types of five-axis machines. Both have five axes, however, the difference is in how they operate. The difference is actually a difference in the control and not in the mechanical configuration of the machine.

The first type of machine allows up to three axes of simultaneous motion. This means that any motion can consist of one to three axes moving at the same time. This allows any specific head orientation to be achieved by rotating the two head axes and then it allows the head to be moved around the trim line using the three linear axes.

At times, however, it is necessary to not only cut along a trim line but to change the head orientation while cutting. For example, it may be required to keep the tool perpendicular to the surface being trimmed. Some of these cuts can be completed by a three- axis simultaneous machine provided all three linear axes are not needed for the trim path. If the trim path can be traced using only two linear axes, then the third axis in a motion can be a rotary head axis. If the trim path can be traced using only one linear axis, then up to two rotary axes can be used in the motion. In many cases, this situation can be improved by carefully orientating the part to minimize the number of axes required to achieve the cut path.

If, however, it is necessary to not only move all three linear axes but also rotate the head during a motion, the three-axis simultaneous type of machine will not work. This is actually fairly common, since the pivot points of the rotary head axes are well behind the tool. As the head axes rotate, the tool traces an arc in space. It requires all three linear axes to keep the tool tip on the part while the head is rotating. If we simply rotate one of the head axes and the head is orientated along one of the linear axes, it is possible to keep the tool tip in place using only two linear axes and the move can be made using only three axes. If, however, the head is not orientated exactly along a linear axis, then even moving one rotary axis will require all three linear axes to keep the tool tip on the cut line. This requires four axes simultaneous motion. If both rotary head axes are moved then it will require all three linear axes regardless of the orientation.

This brings us to the second basic type of five-axis machine, which is one that allows full five axis simultaneous motion. With this type of machine, you are completely free to program any combination of axis motions without concern. The only drawback to five-axis simultaneous capability is that it can require a substantially more expensive control system. Some control companies place a very high premium on full five axis simultaneous controls. The primary market for these controls has been in five-axis metalworking for the aerospace industry. These machines routinely sell for over a million dollars so a more expensive control is not a major drawback. In plastic trimming, however, the more capable control can add substantially to the cost of the machine.

If the price difference is not too high, however, the full five-axis simultaneous machine does offer a higher level of flexibility and capability. It can also simplify programming since it is no longer necessary to keep tract of the number of axes in a motion.

There are some high-end features of five-axis machines that rely on five-axis simultaneous capability. Five-axis tool length compensation, five-axis tool radius compensation and constant tip speed all require five-axis simultaneous capability from the control. We will talk about some of these capabilities later in this book.

From this point forward, I am going to simply assume that the machine we are talking about has full five-axis simultaneous capability. I will try and point out those areas where things might not work with a three-axis simultaneous machine, however, this capability is so ingrained in many important functions that it might be difficult to detail every potential problem.

If you are using or contemplating the purchase of a full five-axis simultaneous machine, you are okay. If you are using or planning to use a three-axis simultaneous machine, you should take special care to determine if the particular program or process you are attempting will actually work. You must also realize that although these machines might be slightly less expensive, they will not perform as well nor will they generally offer the same payback as a full five-axis simultaneous machine.

Chapter 2
CNC Router Design

CNC Router Design

The CNC router has become a major part of several plastic processes. It is used extensively in the industry and yet, it is not understood. I believe that the general level of knowledge and understanding of the design, construction and engineering of the CNC router is relatively low. Unfortunately this low level of understanding has lead to many errors and disappointments. In this chapter I am going to try and help this situation by covering the basics of CNC router design.

The CNC router market is highly competitive. There are literally dozens of companies struggling to sell three-axis machines and the number of companies offering five-axis machines is growing. Many of these companies attempt to convince the customer that their machine is better than their competition. They cite a specific engineering feature claiming that their design is acknowledged as better because of that feature. Or, they simply state that they have a feature, weight for example and hope that the customer simply assumes that more weight is better.

Some companies actually present engineering facts and use these to prove that their design is better. In some of these cases, the facts are true, but other factors may actually be more important. In the end, this is a confusing collage of conflicting claims that few customers can make sense of.

In this chapter we will try and make some sense of the engineering and technology surrounding CNC routers. Unfortunately, presenting this information has one major drawback. It may cause some to believe that they are qualified to judge a CNC router on technical grounds.

Machine tool design is a highly complex science. It is a balancing act. There is no accepted, proper way to design a machine tool. Every decision, every component is a compromise. Each time you select a direction to achieve some benefit, you also give up other benefits and capabilities. In no area of machine design can you gain something without giving up something else.

From this perspective, judging machine design is trying to determine which of the available technologies or design alternatives the machine designer chose. These features also work or blend together. A series of design directions that each taken alone might not be the best, could, working together, provide excellent results. This makes judging a particular machine design by examining the engineering specifications a very difficult task. It is a little like trying to guess what a meal will taste like by looking at the recipe. You may get an idea, but the real test is in the tasting. The fact is that there are probably less than a dozen people in the entire United States that are truly able to judge a machine from its basic design. It is very likely that none of these people are reading this book.

The only really accurate way to judge a machine is to judge the results. How well does the machine actually perform its tasks? How good is the quality of the product it produces? How easy is it to use? How well does it hold up in production? How reliable is it? How long does it last? How easy is it to upgrade or change? What does it cost?

It is safe to assume that if a machine scores well on these questions it is a good design. If it does not, it is not a good design. It is possible for each of us to rather accurately judge the overall design quality of a machine by examining how well it actually works in similar production.

When you focus on the engineering detail of a machine design you are in fact saying "I can tell how well a machine will perform better by looking at the design than by looking at how it actually performs". When we put it this way, it sort of looks ridiculous, yet that is exactly how many companies try and sell their machines and how many companies buy their machines.

Now that we understand this, it can still be useful to understand basic CNC router design so that we understand the decisions that our machine designer made.

We must now understand that a CNC router is not a modified metalworking machine. Many in our industry believe that metalworking machines are much more advanced and much more sophisticated than CNC routers. The opposite is actually true. Certainly the design demands of a CNC router are much more difficult to achieve than a metalworking machine.

Metalworking machines are designed to achieve very tight accuracy tolerances but at a very slow feed speed, typically 1/10 to 1/100 of that required by a CNC router. This difference in operating speed speaks to the core of differences in design between metalworking machines and CNC routers.

As speeds increase the dynamics of everything come into play. It is substantially more difficult to achieve five or ten thousandths of an inch accuracy operating at 100 or 150 inches per minute than to achieve a half thousandth tolerance at 10 inches per minute. Because of its higher speed, a CNC router will travel 10 to 100 times farther during its operating life than a metalworking machine. A combination of weight, slides and friction that offers a more than adequate life for slow moving metalworking applications might prove to have only one tenth to one one-hundredth the life needed for a CNC router application.

In metalworking, the control is simply cooled with a fan and a coarse filter. Metal chips are rather large in size. The factory air might have coolant or oil suspended in it but there is little danger in it condensing on the control components. In many plastics, wood and composite environments, the air is filled with suspended particles. These come in all sizes from very coarse to extremely fine. If these particles are allowed to enter the control enclosure they will be deposited on the control components. This will cause few problems in the first couple of years, but after five or six years problems can be severe enough to consider scrapping out the entire machine. Again, a design decision that is more than adequate for metalworking doesn't work well in industries served by CNC routers.

The design decisions that are made in designing a metalworking machine cannot be the same as those made in designing a CNC router. The requirements are vastly different and require a fundamentally different design. If you try and judge a CNC router design by using criteria from the metalworking industry, you will likely end up with totally misleading results.

In the remainder of this chapter we will focus on specific requirements of CNC routers and compare the different ways these needs can be met.

We have indicated that the design requirements for metalworking machines and CNC routers are different. One basic difference has to do with rigidity and weight. This is an area that generates huge levels of confusion many times fostered by machine manufacturers trying to justify their own design decisions.

Let us return to fundamentals. The purpose of the machine is to produce accurate, high quality parts. High quality generally means smooth even cuts. It is generally believed that this is achieved by using a rigid, accurate machine and at least today this is still true. It very well may not be true in the very near future.

There is a great deal of work being done on machines that react to the machining process. High-speed sensors look at the machining process as it is occurring. The instant it sees the machine reacting to the cutting forces it instantly compensates for the machine movement. In theory, the machine could be made of rubber and still machine very accurate parts. Once this is achieved, machine structure and rigidity will no longer be important. Today, the machine must be more accurate than the most accurate part you want to make with it. In the near future, machines may be able to make parts that are more accurate than the machines they are made on.

An important point is that at least today we are interested in machine rigidity. Not mass, not weight but rigidity. They are not the same thing.

In the past, in the metalworking industry, rigidity was achieved by adding mass. One could accurately assume that the heavier the machine the more rigid the machine. This is no longer true. We are not now in the past and we are not in the metalworking industry so these concepts may not work for us. In fact, applying outdated thinking to modern machines may result in a machine that neither performs well nor lasts very long.

Ultimately, we want a machine that will produce a good quality part and a machine that will last a long time. Today's technology requires a rigid machine to produce a good part. The machine rigidity resists the cutting forces and the dynamics of operation to achieve the cut path desired.

The practice of achieving rigidity through mass that was practiced in the metalworking industry does not work as well in the plastic industry. While you can develop a very rigid machine by making it very massive, this approach sacrifices other requirements. Some of these requirements exist in three dimensional machining and trimming but don't exist in metal.

Cycle times in metalworking are generally fairly long with cycles of ten minutes to an hour being common. Cycle times in plastic processing are much shorter. Feed speeds in the metalworking industry are relatively slow while feed speeds in plastic are fast. Cycle times are heavily influenced by the acceleration and deceleration rates of the machine. The acceleration rate, (how quickly you can reach top speed) and the deceleration rate, (how quickly you can stop), are key factors when feed speeds are high but are much less important at low feed speeds.

The engineering choices made when designing a metalworking machine tend to favor achieving rigidity by adding mass. As a machine becomes more massive, it becomes more difficult to move. It requires more power from the drives and places more loads on the slides and drives. It is more difficult to accelerate and decelerate a massive machine.

These disadvantages are relatively minor in a metalworking machine since neither acceleration nor high top speeds are important. In a plastic trimming application, however, both acceleration and top speed are very important. The disadvantages of achieving rigidity by simply making the machine very massive also make it less desirable for plastic trimming applications.

Designing a machine for trimming applications requires developing a rigid structure that is still light enough to achieve proper accelerations and decelerations. This is a much more difficult task than faced by the metalworking machine designer.

We are not going to endorse any specific method of achieving the design goals. Each machine designer has his or her own approach. Some simply opt for a massive machine, install equally massive drives and oversized rails and drive screws to handle the higher loads and simply live with the higher costs. Others create a simple light structure and tolerate a lower level of rigidity. Still others use computer aided stress and vibration analysis to develop structures that are both rigid and light. There are many variations between these approaches.

It is virtually impossible to try and second-guess the design approach for a machine tool. The only way to tell if a machine design is any good is to watch it operate. Compare cycle times, product quality and machine price and determine from these if the machine is right for your application.

Certain companies in the industry promote the idea that only with a massive machine can you achieve the results you want. They claim that cast iron is superior to other materials for machine tool structures. As we have already shown, neither of these statements is absolutely true.

There is no one superior material. Each material has properties that help and other properties that hurt your design. Cast iron provides mass and rigidity but it is not a particularly strong material and it is subject to cracking with shock loads. Steel provides very high strength and rigidity but must be fabricated into a structure rather than being cast like cast iron or aluminum. Aluminum is rigid and lightweight but has lower yield points than steel or cast iron. Each material has its place in a machine design and claiming that one material is always superior is irresponsible.

Certain exotic materials such as aluminum honeycomb are many times more rigid pound for pound than steel or cast iron. These materials have been used to produce CNC routers for specialty applications but at least today they are much too expensive for widespread use.

At this point, we have examined the basic configuration and basic design requirements for a CNC router. You should have a feeling that there are no right answers. This is actually true. The right machine design and the right machine configuration is the one that fits your application the best.

Drives

Now, let us turn to the drives, the devices that cause the machine to move. Three different drive motors have been used to power CNC routers: stepping motors, DC servo drives and AC servo drives. Again, each has advantages and disadvantages.

A stepping motor uses a permanent magnet rotor and wire wound stator. The stator is wound so that the motor has a large number of poles, typically 200. These poles are generally arranged in groups of four. When one pole is energized, the rotor will align with that pole and is locked in place. The force that it exerts holding that position is essentially the amount of torque available from the motor.

When an adjacent pole is energized and the current pole turned off, the rotor “steps” over to the new position. This stepping motion between poles is the origin of the stepping motor name.

When the third pole is energized it again moves as with the fourth pole. The next step, or the fifth pole uses the same control pattern as the first pole. As we said earlier, the poles are arranged in groups of four. This is an important concept because it may influence performance and accuracy.

When any particular pole is energized, the rotor is locked in position. Should you exert enough force to rotate the rotor from this locked position it will jump over four poles and align with the fifth pole. The control signals that hold the rotor in position repeat every four poles which means that with any position signal the rotor can be locked in any of 50 positions on a 200 pole motor.

The basis for using a stepping motor to run a machine is to take control of the rotor, use a reference, such as a limit switch to determine exact position and rotate the motor carefully enough so that you never lose control of the rotor. If you never skip steps but always keep control of the rotor, the computer will always know the current position of the axis. Since no position feedback is used, this is called an "open loop" system. The control commands a motor position and then simply assumes that the machines follows the command.

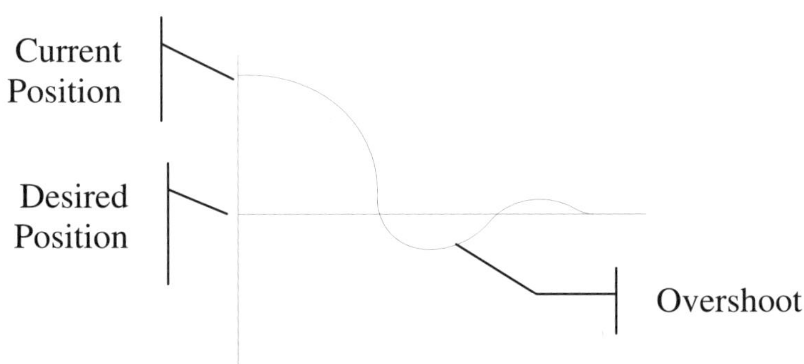

Stepping Motor Step Overshoot

In theory this sounds quite simple, however, large stepping motors have a characteristic that makes controlling them a bit more difficult. Let us examine what happens when you move from one step to another. When the step occurs, the rotor feels a force trying to move it to the next pole. This force causes the rotor to rotate toward the next pole. When it reaches the next pole, the rotor is still moving so it tends to move past it or overshoot until the pull of the pole stops its rotation and starts it moving back. When it again reaches the pole it is again moving, this time in the opposite direction and not quite as fast. It again passes the pole, stops and reverses. Thus when the rotor moves from one pole to another it oscillates about the new pole for a fraction of a second until it finally comes to a stop. This characteristic can be a problem at certain specific speeds. Let's say the rotor moves past the new pole, reverses and moves past the pole again, back toward the original pole. If, at that very instant another step occurs, the rotor is farther from the new pole than it would normally be. Thus the force on the rotor will be greater and it will oscillate farther to each side of the next pole. If again the next step occurs when the pole is farthest away from the new pole, the overshoot again increases. Within a few steps, the forces become high enough that the rotor skips four poles and the control no longer knows the rotor position.

There are several ways of addressing this problem, but for a stepping motor drive to function properly it must be addressed.

The most fundamental approach is to vary the voltage or power sent to the stepping motor based on rotor rotation speed. If you can reduce the force trying to rotate the rotor you can slow down movement from one pole to the next. The idea is to try and balance the time it takes to move from one pole to the next with the time between steps. If you can do this properly, as the rotor reaches the next pole, the system steps again and there is not enough time to oscillate about each pole.

Although the theory is relatively simple, many factors including machine mass and cutting loads affect rotor rotation time with a fixed voltage. This means any system designed to eliminate oscillation must be a compromise and will not work perfectly under all circumstances.

Another method of controlling stepping motors is called microstepping. Instead of simply energizing each pole, one at a time, the microstepper systematically balances forces between two adjacent poles to carefully rotate the field and thus the rotor. This system does pretty well eliminate oscillation but because much of the time the rotor is being acted on by two opposing poles, overall torque may be reduced.

Stepping Motor and Viscous Dampner

Another device that can help eliminate oscillation is a viscous dampener. This is machined canister connected to the rotor shaft of the motor. Inside the canister is a mass which is free to rotate with only a few thousands of an inch clearance from the canister walls. This space is filled with a highly viscous fluid. This arrangement then provides a large amount of rotary dampening primarily in the speed range where oscillation occurs.

Stepping motors were used in early CNC routers primarily because they were substantially lower in cost than other drives at the time. Properly engineered, they perform well in less demanding applications. Again, when properly engineered the open loop feature was not a significant problem although in practice it was a good idea to perform the reference procedure often.

As the cost of DC servo drives and AC servo drives came down, stepping motors lost favor. Their greatest advantage, cost, was no longer as significant and the technical effort required to properly apply them was greater than with the alternative drives.

Today, a number of new, low cost machines have again begun using stepping motors. When evaluating these machines, make certain that they are engineered properly to take care of oscillation and that they are balanced enough that they do not lose steps when executing a program.

The next most common drive is the DC servo. A DC servomotor is configured with a permanent magnet stator and a wound wire rotor. The permanent magnet magnets making up the stator are attached to the outer case of the motor. A set of carbon brushes transfer power through an armature to the wound stator.

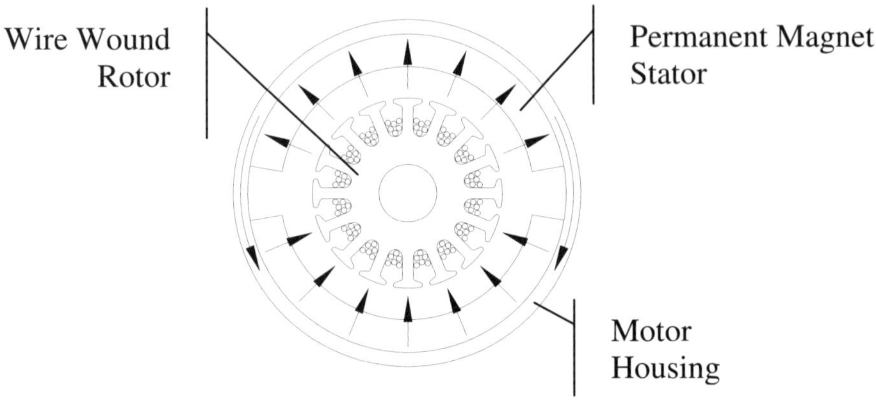

Diagram of DC Servomotor

A position feedback sensor or encoder is attached to the rotor so that as it rotates, a signal is sent to the drive indicating the position of the rotor. When the system is first energized, the axis moves until a switch or other signal determines a reference machine axis position. Once the reference point is achieved, the control keeps track of the rotation of the servomotor through the encoder. This is a closed loop system unlike the stepping motor. Not only does the control dictate the position, but it also checks through the encoder to determine whether or not its control signals have been executed.

An axis controlled by a stepping motor is positioned by feeding the proper number of steps to the stepping motor drive. The motor then steps to the correct position. A DC servo works differently.

As we said earlier, the control knows the position of the DC motor rotor by reading the encoder. When the control wants the motor to move, it knows where the motor is currently positioned and where it should be positioned. The difference between these two positions is called an error signal. This error signal is fed to the servo drive, amplified and fed to the servomotor to cause it to turn in the direction needed to eliminate the error signal. As the error signal becomes smaller, the voltage fed to the drive also becomes smaller and motor rotation slows until there is no error signal and the drive is stopped.

Although there is no force on the motor in the stopped position, as soon as you try to rotate the rotor, an error signal is developed, amplified and fed to the servomotor to resist your motion. It is as if the motor is locked in position.

The amount that the error signal is amplified is called the "gain". The servo closure loop generally takes into account other factors than just the error signal to try and obtain better performance. Sometimes the error signal is mathematically integrated over time. This means that the longer an error signal exists for an axis, the harder the motor tries to close the gap.

To obtain motion of the axis, the control continuously redefines the position that the motor should be at. Error signals are developed as the desired position becomes different than the current position forcing the motor to "chase" the point that was programmed.

The greatest advantage that the DC servo system has over the stepping motor is that there is no danger of losing steps. The closed loop nature of the design means that the current position is always known. Acceleration and deceleration in a stepping motor system must be limited to make certain that the holding torque of the motor is never exceeded.

Otherwise steps and position would be lost. A DC servomotor does not have that same limitation. The torque available for acceleration and deceleration will be higher.

This higher torque generally means better performance. On most part programs, cutting speeds are defined by the material and cutting tool used, and are not a machine limitation. The only difference between machines is generally how fast the machine can accelerate to that cutting speed and how quickly it can come to a stop. These factors can substantially impact cycle time, many times more than top speed.

DC servomotors do, however, have problems of their own that must be addressed. The first centers around that fact that an error signal must be present for the system to work. This means that the machine is never at the position it is supposed to be at. It always lags behind the proper position. The distance that it lags behind is called a “lag error”. If you think about it you will realize that the faster an axis tries to move the greater the lag error.

Another potential problem occurs when you come to a stop. If the axis you are trying to stop is massive and machine friction too small to stop the axis quickly enough, then the servomotor must stop the axis. It must create force in the direction opposite the movement of the axis. To create this force, however, the axis must drive past the desired position and create an error signal in the opposite direction. The actual position is now leading the desired position. The amount that it leads the desired position is called a “lead error”.

Under these circumstances, when the desired position comes to a stop, the axis (which was running ahead of the desired position) will have overshot the stop position. It will need to reverse direction and return to the desired stopping point. If this overshoot occurs while machining an inside corner, a machining defect will result. The only way to address this is to reduce the mass of the axis, or either increase the friction or reduce deceleration so that a lead error does not develop.

Overshoot at part corner

This is prime example of where trying to build a massive machine to achieve high performance doesn't make sense. Mass is the single biggest contributor to degraded performance. I have observed machines in operation where the designer did not take this into account. When cutting square corners, the machine overshot the corner and didn't recover until part way down the next side. It cut a "bubble" where a square corner should have been.

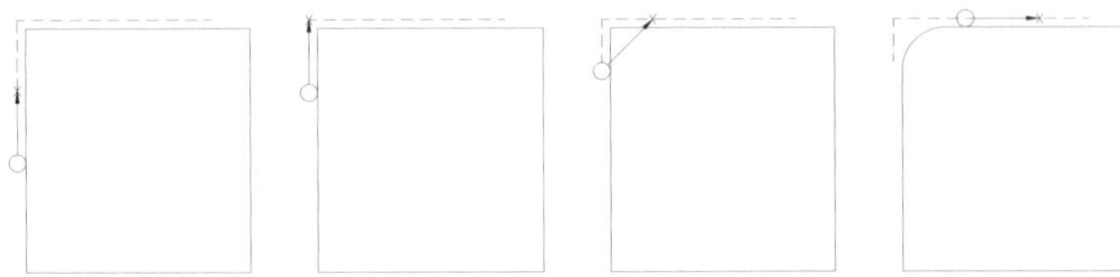

Lag Error Causes Rounded Corner

Another common problem with square corners is rounding the corner. If you think about it, a servo loop is a little like pulling a weight around with a spring. The length of the spring is the lag error and corresponds to the error signal. You trace the desired position with the end you are pulling and the weight follows some distance behind. When you come to a corner, if you do not wait until the error signal becomes zero, you will start pulling the weight sideways before it reaches the corner. Although you traced a perfectly square corner, the weight actually rounded the corner. Many CNC controls actually have a command called Absolute Stop with makes sure that the corner has been properly reached before beginning the next move.

Lead/lag error can also cause problems when you are trying to coordinate more than one axis at a time. Because of lag error, the actual position of an axis is behind the position that was programmed. With a single axis motion this is not much of a problem. Although the position in time is not exactly the same, the path is correct and that is ultimately what you are interested in.

When you try and coordinate the movement of two or more axis simultaneously, you must consider the lag error of each axis independently. The lag errors of each axis must be adjusted so that each axis lag error is the same as all other axes lag error at every speed and every acceleration. This servo tuning procedure is part science and part art.

When properly tuned, the lag errors of all axes balance and the machine travels the programmed path (although slightly behind the desired position in time). To properly tune a machine, however, it is generally necessary to try and balance the mass of each axis to try and get the lag error as close to the same mechanically before you try to tune it electronically. Trying to balance a massive axis with a light axis is almost impossible using a conventional servo closure loop.

Just how large are these lag errors? I have worked with an older set of DC drives, a commercial CNC control and a massive machine where lag errors of four to six inches were experienced. The final machine, however, was used in aerospace and could machine very accurate parts.

Generally, the greater the torque available from the servomotors, the tighter you can make the servo loop and the lower the lag error. Lighter machine will operate with less lag error than heavy machines.

Recently a new type of servo loop has developed that operates with very little lag error. This system can operate with lag errors of only five or ten thousands rather than lag errors measured in inches. This system requires high-speed computer processors in the servo closure loop. These computers generate the error signal, not from the actual error signal, but by looking ahead in the program and calculating what the error signal will be at some point in the future. By combining the actual lag error and the anticipated future error, an error signal is generated that anticipates future movement. If you go back to our analogy of pulling a weight with a spring, this system doesn't lead the front end of the spring around the programmed path. Instead, it moves the lead end of the spring around a path calculated to make the weight move on the desired position along the path. This path is generally calculated to operate with a very small lag error because lag errors generally cause fewer machining problems than lead errors.

For this system to work well, however, requires a responsive servomotor with plenty of torque. How do we get more torque from our servomotor?

We could, of course, install a larger servomotor. While this will give us more torque, to obtain that increased torque requires that wire wound rotor be made larger. This larger rotor has more mass which makes it more difficult to accelerate and decelerate. The mass of the motor rotor is the single biggest factor in determining acceleration.

What would happen if we just fed more power into the existing DC motor? Since the power must be transferred to the rotor through the brushes at some level of power, the brushes will no longer be able to carry the current and will begin to arc. This is obviously the maximum power that can be transferred. If we operate just below this level, however, another problem develops.

As electric power moves through the windings in the rotor, heat is generated. This heat must exit from the rotor across an air gap to the permanent magnet stator, through the magnets to the motor housing and then through the motor housing to the surrounding air. This is a rather long heat path. The efficiency of this heat path determines the maximum average power that a DC servomotor can generate. If the heat being generated cannot be dissipated, the motor will burn up.

To address these problems, a new type of servomotor has become popular today. This motor is commonly called an AC servo. It is also called a brushless DC servo by some companies.

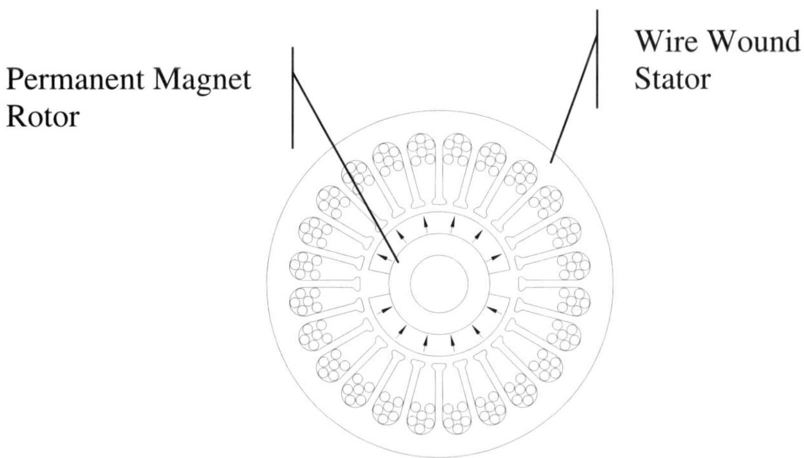

Diagram of AC Servomotor

This motor is constructed differently from the DC servomotor. In an AC servomotor, the rotor is a permanent magnet and the stator is wire wound. Because there are no wires in the rotor, there is no need for brushes. Commutation is performed electronically by the servo drive rather than being performed by the brushes and commutator.

Lets look at the advantages of this arrangement. First, because there are no brushes to arc, the maximum limit to the power you can feed to the motor is only limited to the power required to melt the wires in the stator. As you can imagine, this is substantially greater than the power you can feed through rotating brushes. The average power that a motor can operate at is also higher. Since the wire windings which generate the heat are the stator, they are attached directly to the motor housing. Instead of the long heat path of the DC servo motor, the heat from an AC servomotor must only travel from the windings through the motor housing to the outside air.

The result is that in the same frame size, an AC servomotor will provide more power. It can also generate bursts of power well beyond normal as long as the average power is below the design limit. The permanent magnet rotor generally has less mass than a wire wound rotor making acceleration and deceleration faster. Higher end AC servomotors use rare earth magnets that have less mass and can contain higher magnetic fields that iron cores. These provide even higher acceleration deceleration performance.

These are the three most commonly used drive means for CNC routers. In the early days, some CNC routers were hydraulically powered, however, this method is not used today. Each of the methods can work well, subject to their design limitations, provided they are properly engineered. If poorly engineered, all will perform poorly.

Stepping motors were commonly used because of their low cost. Today, however, the high end of the stepping motor technology and the low end of DC servomotor technology have met. Low-end stepping motors that do not address step resonance are generally not suitable for powering CNC routers. DC servomotors have come down in price and today, these appear to be a better choice for low cost machines.

Almost everyone agrees that AC servomotors have the performance edge, although they generally cost more. You must make sure that the machine design can take advantage of the higher performance drives. It makes little sense to pay more for drives that do not provide additional machine performance.

Now that we have looked at the drive motors, let's examine exactly how they move the axis. There are several methods that are used to convert the rotary motion of the drive motor into linear motion of the axis. Just as before, each system works but each is a compromise.

We will start with two methods that work but are seldom used today. The first is to move the axis with a belt or chain. The drive motor turns a pulley or sprocket that the belt or chain is wrapped around. The ends of the belt are attached to the movable part of the axis.

This system can generate extremely fast movement but it is difficult to achieve the rigid level of positioning needed for router work. Nonetheless, I have seen some very low cost machines use this system and there is no reason why it can't perform well in less demanding applications.

The next system is one I have not seen in the U.S. but is used on machine tools in Russia. It is called a friction drive and consists of a polished steel bar on which a steel wheel rides. The drive wheel is pressed into the bar with a substantial force, and friction between the bar and the wheel is used to move the axis. The drive provides backlash free operation but has some other disadvantages.

Because the drive relies on friction and not a mechanical link like a gear, it can drift over time. To prevent this, they use a linear feedback system along the axis to determine position rather than the encoder on the servomotor. The CNC control must then resolve any conflict in position data from the two sources, that is the linear scale and the encoder on the servomotor. In Russia, they used lasers as the linear feedback, but they were limited to feed speeds that were too slow for woodworking. The system could work with glass scales, however, this type of linear feedback is quite expensive and is difficult to cost justify today.

Rack and Pinion Drive

The two most common drives today are the rack and the ball screw. The rack consists of a long steel bar into which gear teeth have been cut. The drive motor turns a pinion gear that is meshed into this rack. As the gear turns, it rolls up and down the rack moving the axis. Although this system is rather simple it has some limitations.

The servomotor must turn at a reasonably high speed to generate the required power. It must therefore drive the pinion gear through a gear reduction box. For a gear of any kind to operate, there must be some clearance between the teeth. When gears are properly meshed together, they do not slide on each other. They roll on each other kind of like a rocking chair rolls on the floor. There must be some clearance between the backside of the gear tooth and the next gear to prevent these from sliding or scraping on each other. This clearance is called backlash. Some level of backlash must be present for pinion gears to operate. The gearbox may also have some backlash depending on the design.

Backlash can create some difficulty when operating a CNC router. The basic effect of backlash occurs when an axis reverses direction. The drive is moving the axis in one direction. When it stops, the gears are all solid in the direction that the axis was driving. When the drive reverses, it must first take up the backlash of the system before it begins to move the axis. For example, let us assume that the backlash was ten thousants of an inch. When the axis reverses, it must move ten thousants of an inch before the axis begins to move at all. Once the motion is complete, the axis will be ten thousants of an inch short of the desired position because of the backlash.

The simplest way to adjust for this problem is by using backlash compensation in the CNC control. All this does is add the amount of backlash any time a drive changes direction. In the simple example, the reverse motion would be the distance programmed plus ten thousandths. In actual practice, the amount of backlash compensation actually put in the control is slightly less than the actual backlash, but it reduces the error to such a small number that it doesn't matter.

Excessive backlash can create other problems that can't be corrected by backlash compensation. If the machine is operated to cause a lead error, the backlash may create a positional error even with compensation. This is especially true if the next motion is on the second axis.

Also, should the cutting forces become very large with a low friction machine, it can cause the axis to vibrate within the backlash, thereby creating a poor quality cut.

Although most machine manufacturers use one technique or another to try and eliminate backlash, the final goal is the same. If backlash can be reduced to five thousandths or less, it is unlikely that it will have a noticeable effect on either performance or accuracy.

There are also methods for mechanically eliminating lash in a rack system. One technique uses a split pinion gear. The gear is split vertically into two gears and these are spring loaded so that the teeth rotate apart. One gear contacts the front teeth of the rack and the other contacts the back teeth. Although this does eliminate backlash it does so by sliding one of the gears on the teeth resulting in substantially higher wear.

Another method uses two pinion gears, each driven by a different servomotor. One servomotor is controlled normally and the second is controlled to maintain a certain amount of force against the first motor. Again the teeth of one motor will be sliding instead of rolling and in this case a second drive system is required on each axis.

Rack drive systems have some other shortcomings. The gearbox required to drive the pinion gear is generally not more than 75% efficient and in many cases less. This means that a significant amount of the available power is used to overcome friction in the system. If you loosen the gears up to decrease friction you introduce additional backlash. If you tighten the system to eliminate backlash you increase friction and decrease performance.

Because of these problems, the most common drive means for CNC routers is the ball screw. We have already discussed the ball screw design when we covered lead screw compensation. The lead screw compensation discussion also applies to screw drives.

A ball screw is about 98% efficient so it does not have the friction problem that the gear reduction drive has with a rack drive system. A ball screw will generally have some level of backlash, although it is possible to minimize it by selectively using different sized balls that exactly match the screw being used. Backlash can also be eliminated on a ball screw by using a second ball nut spring loaded against the first one. Unlike the spring-loaded system on the rack, this system does not cause sliding. The balls on both ball nuts roll properly so this arrangement does not significantly increase wear.

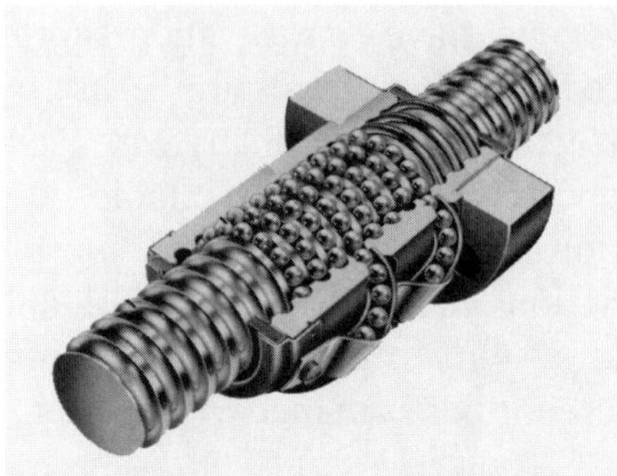

A Ball Screw System

Although lead screws seem to have the edge as a drive means today, they do have one major technical problem that must be addressed. This problem is commonly referred to as lead screw whip. If you take a long metal rod and begin to rotate it faster and faster, at some point you reach a rotation speed called critical speed. Any rotation speed above critical speed will cause the rod to wobble or whip back and forth. Operation below that speed will be steady and operation above that speed will be increasingly unstable.

Lead screws are, in fact, long metal rods and so they are affected by this tendency to whip above certain speeds.

Several factors affect critical speed. The rigidity or stiffness of the material affects critical speed. The stiffer the material that the screw is made of, the higher the critical speed. In the case of lead screws, however, they are all made of essentially hardened steel and so finding a material that is significantly stiffer is difficult.

The mounting method, (the method by which the ends of the lead screw are mounted) also affects critical speed. Without getting into a great deal of engineering detail, the more rigid the end mounting, the higher the critical speed. This is an area where machine manufacturers can improve the situation although rigid multi-bearing mounting assemblies are more expensive.

If you examine a typical lead screw on a CNC router you will find that the lead, (the amount that the screw advances the axis per revolution) is typically one inch. With a one-inch lead and feed speeds normally found on CNC routers today, the lead screw is well into critical speed.

One method of addressing the whip problem is to increase the lead of the screw. In this case, each revolution of the screw moves the axis two or three inches instead of one inch. Thus, the required feed speeds can be obtained with lower rotation speed of the lead screw. These types of screws are typically called high helix screws.

There are some shortcomings with these screws, however. Since they rotate slower, some means of gear or belt reduction is needed to allow the servomotor to operate at the higher rotational speeds where it is efficient and yet drive the lead screw at a slower speed. This reduction can be a source of design problems and perhaps backlash if it is not engineered properly.

Because of their geometry, high helix screws are generally not constructed using the same methods as the more typical half inch or one inch lead screws. This different construction method tends to make high helix screws more expensive. It also tends to give them less overall thrust capacity. Again, the thrust capacity of most lead screws today is many times greater than that required to operate the machine so don't let a fast talking machine salesman scare you about lead screw capacity. If the machine performs well in production and exhibits a good long operating life the lead screw choice made by the machine manufacturer was a good one. When we discuss some of these design factors we are simply trying to show you why some manufacturers design their machines one way and others select a different set of design parameters.

Maintaining accuracy with high helix screws is more difficult. A rotational error on a high helix screw will have two or three times the impact of that same rotational error on a standard lead screw.

Despite their higher cost, high helix screws are successfully used in a variety of applications and are relatively common on point to point machines where very high feed speeds are common. Again, there is an engineering compromise that must be made between screw cost, accuracy, motor rotation speed, power and screw whip.

Another way to address screw whip is to not rotate the screw. By rigidly fixing the ends of the screw and rotating the ball nut instead, critical speed can be increased. There still is a critical speed but it tends to be higher than when the screw is rotated. If you don't rotate the screw, why should there be a critical speed at all?

In practice, the interaction of the ball nut and screw at certain speeds will induce screw whip even if the screw is not turning. The factors that influence screw whip are closely associated with the natural harmonic frequency of the screw. This is the frequency at which the screw will vibrate if hit. This is somewhat like a guitar string vibrates when it is plucked. When screw operation inputs forces at or near this natural frequency, the screw can begin to vibrate,

Each cycle the vibration gets greater until the screw is whipping. This is something like a person walking over a long footbridge. Sometimes the frequency of the steps matches the frequency of the bridge and the bridge begins to bounce. Each step makes the bounce higher. This same phenomenon occurs with a fixed lead screw and a ball nut.

The most common way to address lead screw whip is to use a larger diameter lead screw. For a specific length, the larger the diameter of the lead screw the higher the critical speed. To prevent screw whip completely, simply make the lead screw diameter large enough so that the critical speed is higher than any possible operating speed. This does work well for eliminating lead screw whip, but as always, there is another price to pay. The obvious price is a more expensive lead screw but it goes beyond this.

When we discussed drive motors we focused carefully on obtaining the best acceleration and deceleration possible. We also noted that the mass of the rotor of the drive motor was a major factor in performance. After the inertia of the motor rotor, the rotary inertia of the lead screw is the next most important factor in overall system performance. The larger in diameter the lead screw, the more difficult it is to start and stop it from rotating. The inertia increases with the square of the diameter. If a lead screw is twice as large it does not have twice, but four times the inertia. It is four times harder to get moving and four times harder to stop than the smaller lead screw.

We can resort to a larger servomotor to power the larger screw but that costs more and requires a larger rotor that reduces performance from the motor. A larger lead screw, while it does reduce or eliminate whip generally provides somewhat lower overall performance than the smaller diameter lead screw.

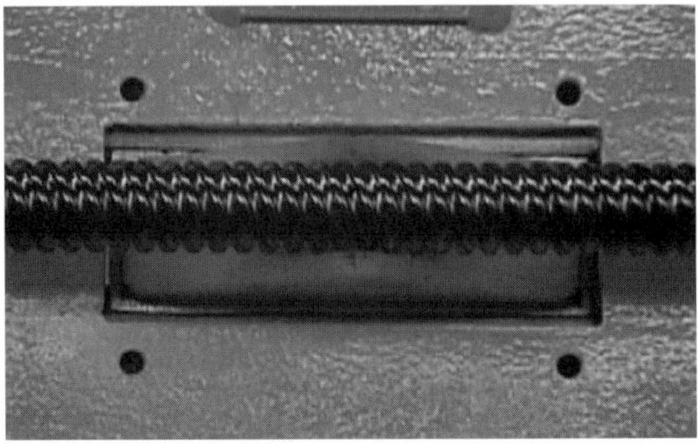

Thermwood Anti-whip System

Thermwood has developed and patented a system that allows a smaller, lower inertial lead screw to operate above critical speed without screw whip. The device is simply a nylon guide with a C shaped cutout. The lead screw fits into the cutout. The device is positioned to support the lead screw somewhere near the middle. The lead screw turns inside the guide. The guide prevents the lead screw from moving back and forth in any plane and thus prevents it from setting up whip. It is a little like placing a bearing in the center of the lead screw to support it. The lead screw then operates as two shorter lead screws each with a much higher critical speed.

When the axis needs to move from one side of the guide to the other it simply pushes the guide out of the way as it goes by. Once the axis mechanism is clear of the guide, the guide moves back into position to again support the lead screw. While the axis is passing the guide, the ball nut acts as the support guide so the anti-whip guide isn't needed.

This system works well, is relatively inexpensive and allows a smaller diameter high-performance screw to operate well above critical speed without encountering problems with screw whip.

Again, when considering the type of drive, each of the systems discussed can be made to work and work well. It is more important that you determine that the machine designers who developed the machine you are considering properly understand the underlying technology. If they do, they will properly address the shortcomings and limitations of the method they have chosen and the system will perform well. Again it is important to look at actual system performance rather than try and determine theoretically which is the better system.

The next item we will discuss is the linear slides or rails. Again, there are several methods that all work reasonably well, and again, each has advantages and shortcomings. One or two of these systems are used extensively, however, each has been used on CNC routers in the past, and some may show up again in the future.

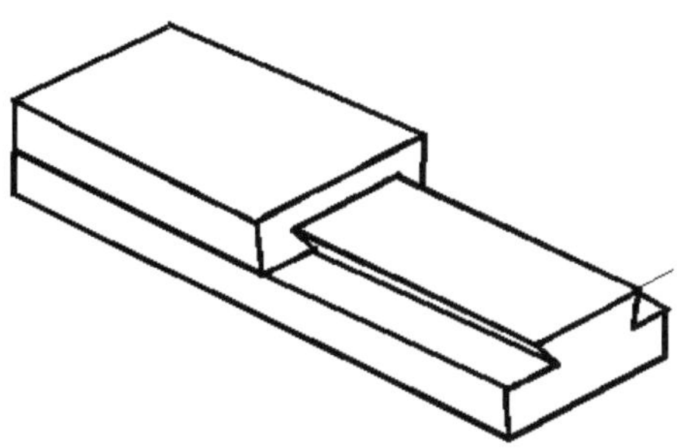

Dovetail Slide

There are some systems used in the metalworking industry that are not normally found in industries served by CNC routers. The simple dovetail slide is one of those systems. The slide is made up of a ground and polished metal slide, flat on top with edges that turn back like a dovetail. The carrier is machined to fit this dovetail, generally with an adjustable bar along one edge. This carrier simply slides along the polished slide relying on some type of lubrication to reduce sliding friction. This system offers a highly rigid slide with very high load carrying capacity. Some of the shortcomings of this system, however, are reducing its use, even in metalworking machines.

The system exhibits a high level of friction, especially when loaded. With proper lubrication this is not a problem, especially at lower speeds. As feed speeds increase the friction, besides reducing system performance, also generates heat. This heat can become excessive in high feed speed applications and bind the carrier and slide in extreme instances.

The need for a constant level of lubrication makes it difficult to operate in a wood environment. Sawdust that collects on the slide can absorb the lubricant and carry it away as the sawdust is pushed off the slide during operation.

The final shortcoming for CNC router applications is a life that is too short at the feed speeds common in woodworking operations. The high friction tends to wear the slide. The slide can be adjusted for this wear to some extent, however, it is common for the wear to be greater near the center of the slide where the carrier moves most of the time and less at the ends. If you adjust the slide to be tight in the center, it won't move to the ends because it jams. If you adjust it properly at the ends, the center is too loose. At this point, the slide must be rebuilt or replaced.

A variation of this system is used in large metalworking machines and on some axes in the woodworking and plastics industries. This system substitutes a polymer material on the slide carrier for the metal gibes. In addition to the other advantages, that is a rigid structure and high weight capacity, this system offers much lower friction and lower wear than the metal slide.

There are two types of polymer used for this purpose, thermoset and thermoplastic. A thermoset is a liquid plastic that undergoes a chemical reaction when processed and turns into a solid. Some common thermosets are polyester in fiberglass and epoxy (commonly used in two part adhesives) Once a thermoset liquid has reacted and turned solid, it will not turn back to a liquid again.

Thermoplastics are plastic materials that soften when heated above a certain point and harden again when cooled below that point. Each time a thermoplastic is heated it will soften.

Because of their lower friction levels, polymer slides can handle higher average feed speeds than metal slides. They have been successfully used on vertical axes of CNC routers where the high rigidity is important, but we are not aware of this material being used on any of the long major axes.

Some of these materials are impregnated with lubricants and can operate without external lubrication. Some of these materials were developed for the space program for use in space where liquid lubricants freeze. This feature can eliminate some of the problems associated with liquid lubricants in dusty environments such as wood or fiberglass.

Application of a polymer slide material is actually quite complex. There are two types of friction, static and dynamic. Static friction is the friction that exists between two surfaces that are stopped and resting against each other. It is the friction that resists sliding one object against the other from a stop. Dynamic friction is the friction that exists between two surfaces once they are moving against each other. It is the constant resistance as you slide one surface against another.

Static friction is generally greater than dynamic friction. You must push harder on an object to break it loose and get it moving. Once moving it requires less force to keep it moving. We have come to call this characteristic stickshion.

Material that exhibits stickshion is difficult to control using a servo system and should be avoided. There are some materials, however, whose static friction and dynamic friction are almost identical. These materials are ideal for use on machine tool slides and in fact, they are used extensively for that purpose in metalworking.

Another approach that further eliminates friction is to use a steel bar and recirculating roller bearings. This system offers very high load capacity and good rigid construction, however, it only provides support in one direction. To develop the all around support needed, several roller bearings are required sometimes opposing each other. Although this system works well, it tends to be more expensive than required and is only used today on heavier machines where the load capacity is needed to carry the weight of the moving components.

Round Way and Round Way Linear Bearing

On the other end of the scale is the round way and linear bearing. The round way is a ground, polished rod on which rides a recirculating ball bearing. Round ways come in a variety of diameters and load capacities. It offers two major advantages. It is low cost and it has very low friction.

Because the round way and linear bearing is actually a round ball riding on a round shaft, the contact between the ball and the shaft is a single point. Even though there are a number of balls to carry the load, the load carrying capacity of this system tends to be lower than the other systems we will discuss.

The ball nut wraps around about three-quarters of the way around the shaft and is open at the bottom to allow clearance for the way support. This construction method means that the load carrying capacity of the system is different depending on the direction of the applied forces. Pushing toward the support rail offers full load capacity. Forces perpendicular to the support rail have a reduced capacity, typically 70% of rated load. Forces pulling away from the support rail may see the load capacity cut almost in half.

This characteristic must be taken into account when the fundamental design of the machine is first developed. It means that either the designer must configure the axes so that forces are applied in the optimum direction or must specify larger round ways that can handle the loads in a less than optimum configuration.

Track way and Track way Linear Bearing

Today, the most popular slide system is the track way. A track way is essentially a steel bar into which has been machined four concave tracks. A recirculating set of ball bearings rides in each of these tracks.

The track way has several advantages over the round way. Because you have a convex ball riding in a concave track, contact between the ball and the track is a line rather than a point. This means that the load capacity of each ball is greater and therefore, the load carrying capacity of a track way is several times greater than the load capacity of an equal sized round way. Also, because of the opposing orientation of the four tracks, the load capacity of the track way bearing is essentially the same regardless of the direction that the force is applied. This tends to make the track way more rigid overall than a similar sized round way.

The only real disadvantage of the track way is that it tends to cost more than a similarly sized round way.

Again, the machine designer is faced with a series of selections for the machine ways. Applied properly, each will perform well. If you ignore the limitations, however, problems can occur. Again, the best way to determine if the design is adequate is to see how systems in the field are performing and how well they hold up over time.

Positioning of the rails has become a point of discussion by some CNC router companies that claim that placing one track on top of the support tube and one track on the front face is fundamentally superior to placing both tracks on the front face.

Since a good quality track way has essentially equal load carrying capacity in all directions, and properly designed, the rails and bearing sizes selected should have substantial overcapacity, there does not seem to be a real basis for this claim.

My company built machines using this scheme and round ways in the past. In this case it did orient the round ways so that its maximum load-carrying capacity was vertical. We eventually abandoned this arrangement because we found it extremely difficult to achieve proper alignment of the two rails mounted on two perpendicular surfaces.

If rail alignment can be achieved, this system should work just fine but I sincerely doubt that there is any advantage to it using modern track ways and bearings.

The next area we are going to discuss is the router spindle. The technology in this area is quite complex and there is a lot of misinformation about spindles in the industry. Essentially there are two different types of spindles in use, the belt driven stand-alone spindle and the motor spindle.

The earliest spindles used for routing were the belt driven spindles. They consisted of a spindle housing with bearings at both ends that supported the spindle shaft. On one end of the spindle shaft was a collet or other arrangement to hold the router bit and on the other end was a pulley through which the spindle was driven. To obtain high router speeds, the electric drive motor was equipped with a large diameter pulley and the spindle shaft was equipped with a small diameter pulley. If the motor pulley was four times larger in diameter than the spindle pulley, the spindle would rotate four times faster than the motor.

The routing process can consume several horsepower, so the drive belt must be capable of transmitting the necessary horsepower. The early spindles used wide flat belts running on flat pulleys. These were capable of transmitting the required horsepower between the drive motors and the spindle.

This type of spindle was primarily used on fixed routers, such as pin routers. The flat belt driven spindle was used successfully on early CNC routers although the package was large and somewhat heavy. Today the flat belt driven spindle is not used on CNC routers (although Thermwood has developed a variation that is used extensively).

In the fundamental design, the Thermwood spindle makes two basic changes from the original flat belt driven spindle. First, it does not use a flat belt, but instead uses a modern design poly-V belt. This belt uses a set of formed ridges along the belt to increase surface area and drive capability. This means that a belt that is less than an inch wide can transmit ten horsepower where it would require a five or six inch wide flat belt to transmit the same power. The second fundamental change to the design is that the driven pulley is located at the bottom of the spindle shaft, just above the collet instead of on the opposite end of the spindle. Thermwood engineers believe that you obtain a better drive with less spindle stress by driving the spindle as close to the cutting area as possible. Although substantially smaller than the original flat belt driven spindles, the Thermwood spindle is still much larger than the same horsepower motor spindle.

One potential problem with driving the spindle near the cutting area is that the pulleys and belts are subject to dust and contamination. Without effective protection, sawdust will build up on the pulleys, increase the tension on the belt and possibly destroy the balance of the motor or the spindle.

To prevent this from occurring, Thermwood runs both the drive pulley and the spindle pulley inside a closed chamber. The chamber is pressurized with compressed air to provide a small positive pressure inside the chamber. Any possible hole through which sawdust could enter has air leaking out of it. This effectively prevents anything from entering, even in highly contaminated environments.

As CNC routers became more popular, demand developed for a router spindle that was smaller and lighter than the belt driven spindles of the day. It was determined that a synchronous motor wound to rotate at 3,600 RPM on 60 cycle electric current would turn at 18,000 RPM if operated on 300 cycle current. To obtain the 300 Hz frequency, engineers took a standard 60Hz electric motor operating at 3600 RPM and coupled it directly to a generator that produced 300Hz current when turned at 3,600 RPM. The result was current that would drive the motor at 18,000 RPM.

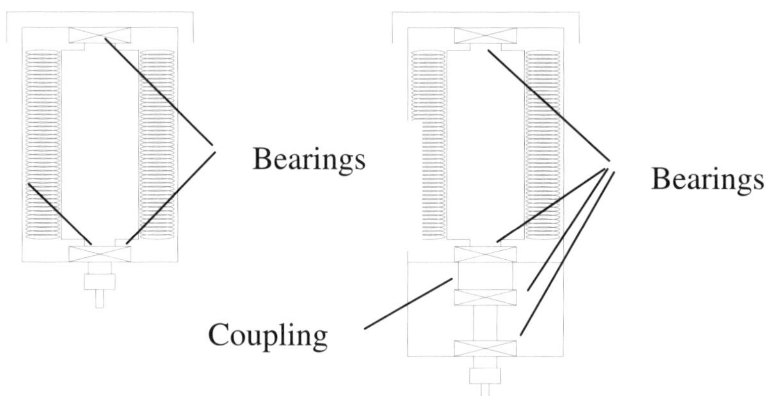

Spindle-Motor **Motor Spindle**

To turn this motor into a router spindle they simply added a collet arrangement at one end to hold the router bit. Although it worked, problems quickly developed. The typical spindle motor had a single ball bearing at each end designed to handle the rotational forces of the motor. These bearings now needed to handle not only the motor loads but also the cutting forces generated by the routing process. Also, heat generated by the routing process was introduced directly to the motor shaft exposing the bearings to the heat. With higher loads, faster speeds and higher heat, bearing life was reduced drastically.

To address this situation, a separate spindle with its own bearings was coupled in-line with the motor. This arrangement became known as a motor-spindle. In this arrangement the motor had two bearings and the spindle also had two bearings. The spindle bearings could now handle the cutting forces while the motor bearings were again only required to handle the motor loads. This arrangement worked better in most applications but some other problems developed.

The bearings used for the spindle are an angular contact ball bearing. This means that the balls of the bearing contact the bearing race at an angle. A router bit is subject to two forces. The first is the radial load. As you push the bit through the work an opposing force pushing against the bit is developed. The primary purpose of the bearing is to resist this force.

As you push on the bottom of the tool, straight into the spindle, you develop the second force, axial force. This is the type of load developed when you plunge the router bit into the workpiece. Also, certain tools, especially shaped cutters, create both radial loads and axial loads.

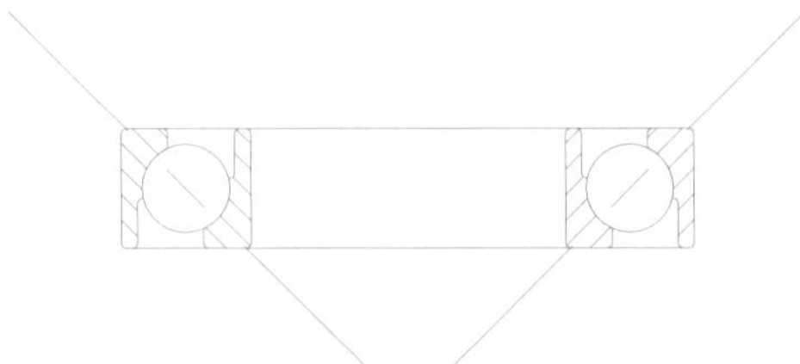

Angular Contact Ball Bearing

The contact angle of the angular contact ball bearing determines the ratio of radial force to axial force that the bearing can withstand. A zero angle bearing will have all the bearing capacity in the radial direction but will have little axial capacity. As the angle increases, the axial capacity increases but the radial capacity decreases. Determining the correct bearing angle for a spindle is a difficult decision. In most applications the designers have specified 90% or so of the bearing capacity for radial loads assuming that typical applications are primarily radial. When an application occurs where more axial loads are generated, the bearings fail early.

One common cause of bearing failure from excessive axial loads is trying to plunge into the workpiece with a router bit that is not designed as a plunge bit. Some tools are intended to plunge directly into the work, however, most are not. Using a non-plunge bit to plunge into the work will generate substantial axial loads and will quickly fail bearings designed primarily for radial loads.

One answer for this is to add a bearing at the bottom of the spindle that is primarily intended to handle the axial loads along with the standard bearing intended to handle radial loads. This two bearing set can resist both axial and radial loads.

Companies that build both the spindle and the machine can coordinate the design of the spindle to the drive forces of the machine. This means that the axial capacity of the spindle can be made greater than the ability of the machine to push down. In this case, the machine is incapable of destroying the bearings through excessive axial loads.

A second problem that has occurred with the motor-spindle is in the area of the coupling between the motor and the spindle. In this area we are trying to transmit high horsepower at a very high speed. If we rigidly couple these two, we will set up vibration and forces that will quickly destroy the adjoining bearings. Any misalignment between the motor and spindle will also put stress on the bearings, quickly destroying them. Therefore, some type of flexible coupling is required. Most standard flexible couplings can't operate at normal routing speeds.

The method that major manufacturers settled on is to cut a keyway in the top of the spindle shaft and a matching keyway in the bottom of the motor shaft. A metal key is then inserted in the keyways connecting the shafts. When the shafts are aligned both axially and angular, this arrangement seems to work well. Any misalignment, however, results in the metal key sliding against the shaft 18,000 times a minute. This creates heat that is transmitted up and down the two shafts and ultimately to the bearings.

Instead of resorting to a separate spindle, some manufacturers simply installed a double bearing set on the bottom of a spindle motor. This generally requires larger diameter bearings to fit over the taper of the collet, which restricts the maximum bit shank diameter that the spindle can accommodate. Except for this limitation, the system seems to work reasonably well.

Another factor that seems to have a major impact on bearing life is the fit of the bearing into the bearing seat. The bearing seat is machined into the spindle housing and is generally slightly smaller than the outer diameter of the bearing. The bearing is then pressed into this seat. The press fit actually shrinks the outer race of the bearing slightly so prior to pressing into place, the bearing must be made slightly larger than needed. This fit is extremely important and is quite difficult to machine accurately.

The axial alignment of the bearing seats is also extremely important. If the bearings are slightly out of alignment they will be under a load even before any external cuttings loads are applied. A slight error in seat diameter and a slight error in alignment can easily place the bearings under enough load to use up half or more of the total bearing capacity. A bearing in this condition will fail early compared to a bearing properly installed.

Thermwood has developed and patented a simple system to assure proper bearing alignment on its belt driven spindles. We use a bearing that has the proper fit between the inner and outer race without compression from pressing it into a seat. Originally, the bearing seats are machined substantially oversized and then the bearings are chemically bonded into place using a special polymeric material. Since the polymer is a liquid that sets into a solid, it can exert no force on the bearing during the assembly process. The result was a perfectly aligned bearing with no assembly stresses.

Recently, Thermwood began mounting the bearings in a machined bearing seat that was then bonded into a seat machined into the spindle housing. This has the advantage that the bearings can be replaced without the need to machine away the bonding agent and re-bond the bearings. This makes field replacement of bearings possible.

Most router spindles use standard bearings with ground and polished balls. The steel ball offers a hard, rigid surface that is also somewhat elastic. The hard surface provides a long operating life at high speed and the resilient nature of steel offers a bearing that can handle shock loads and slight levels of contamination.

These standard bearings seem to work well for high-speed applications until the speed reaches about 30,000 RPM. As the bearing speed increases beyond this point, the centrifugal force of the steel balls spinning around the race start to become significant. As the bearing spins, the balls are trying to fly out of the bearing but are held in place by the outer race. The rather heavy weight of the balls can create forces at high speeds that approach the capacity of the bearing. This leaves very little capacity for handling the actual cutting loads.

For very high-speed applications, typically 30,000 to 80,000 RPM, a different type of bearing is used. In this new bearing, the inner and outer bearing races are still made of steel, but the balls are made of a ceramic material. These ceramic balls have a very hard outer surface yet they are very light-weight. Their light weight means that at high speeds they do not develop near the internal forces of a steel ball. The amount of force developed is directional proportional to the weight of the balls.

Using ceramic balls we can again offer a reasonable load capacity at very high speed operation. There is, however some downside to using ceramic bearings. Unlike the steel balls, ceramic balls are not as resilient. They are somewhat brittle and can shatter under a shock load. They are less tolerant of slight contamination and are generally not as rugged.

In general, for speeds up to about 30,000 RPM, standard steel balls provide the best combination of capacity, life and resilience. For speeds above 30,000 RPM ceramic bearings will be required, but the resulting bearing will be less tolerant of misuse or poorly balanced tooling.

Ceramic bearings have one other characteristic that can be useful. Heat is one problem that can quickly destroy a high-speed bearing. Heat causes a problem because as the bearing heats up, the inner race and the outer race both expand. If the bearing is heated evenly, they will expand proportionally so that the race in which the balls ride will remain approximately the same size. The problem occurs when the balls heat up and expand. The expanded balls no longer fit the bearing race. They preload into the race reducing the capacity of the bearing. The preload also increases the heat generated by the bearing itself and before long the bearing fails.

Ceramic balls do not expand very much with heat. This means that in a higher heat application, they will generally last longer than standard steel balls.

Bearing lubrication is an important part of spindle life. There are two generally accepted methods of lubricating high-speed bearings, grease and oil. The grease used for high-speed spindle bearings is a special material. Normal lubricating grease will burn out spindle bearings in well under an hour. Grease bearings are available in two forms. One bearing is sealed and greased for life. The other type is designed to be re-lubricated at regular intervals.

In general, the type of bearing that is re-lubricated will offer a better life than permanently lubricated bearings, provided the lubrication process is done properly. Over many years we have determined that this is almost never done properly in most factories. There are many errors that can be made and any of these errors will result in premature bearing failure.

The most common error is over lubricating. The bearings are designed to operate with very little grease. If too much grease is introduced into the bearing, it will quickly heat up and fail. If the grease zert is not properly cleaned, dirt on the zert will be introduced into the bearing and it will fail quickly. Incorrect grease will cause early failure.

Although the re-lubricated bearing has the potential for the better life, in actual practice the permanently lubricated bearing exhibits better real life in a typical factory environment.

The second basic lubrication material is oil. The oil is generally introduced into the bearing as an air-oil mist. This mist is a fine spray of compressed air into which has been suspended fine particles of oil. This type of lubrication offers several advantages. First, since a fine stream of air is being introduced into the bearing, the air escaping every opening in the bearing effectively blocks contamination from entering the bearing. The fine oil mist provides a minimum level of lubrication but does not build up and cause heat. The air and oil moving through the bearing actually provides forced cooling that results in better bearing life.

Overall, air-oil lubrication offers the best lubrication method but there is a price to pay. The systems required to create the mist and then transfer it to the bearings is expensive and requires ongoing maintenance. A failure of these systems will result in a bearing failure in short order. The air used to power the lubrication system must be dry and free of contamination or, again bearing failure will result. You must also make certain that the exhausting air mist does not condense oil on any part of the machine or spindle where it can drip on the workpiece and ruin it. Some spindles actually try and condense the oil and offer systems to collect and dispose of this material.

As with all other aspects of CNC router design, either lubrication system can work well if properly designed. Again proper design can only be judged by looking at operating results in typical production environments.

When high frequency spindles were first introduced, they were powered by a motor-generator set as we described earlier. The electric output of a motor-generator is a perfect sine wave. The spindle motor is designed to operate with a perfect sine wave. Eventually, electronic frequency converters began to replace the more expensive motor-generators for powering high-frequency spindle motors. With the electronic frequency converters came some new problems.

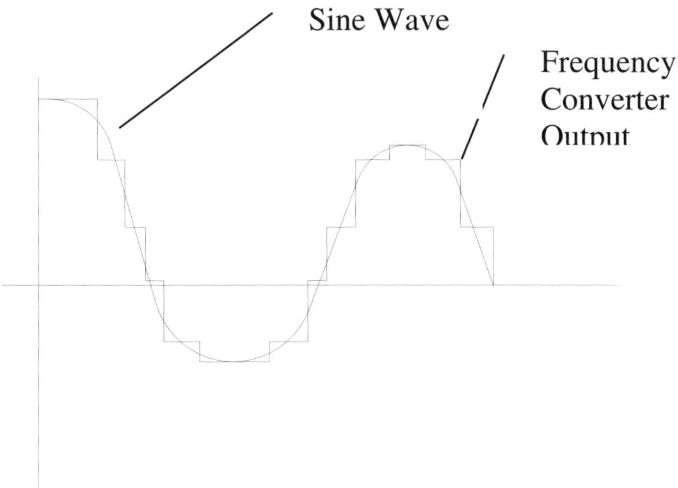

Sine Wave versus Square Wave Frequency Converter Output

An electronic frequency converter is basically a digital device, so producing a smooth sine wave is quite difficult. The electric output of an electronic frequency converter is therefore a modified square wave. This wave shape approximates a sine wave, however, there are areas where the wave form varies from a true sine wave. Wherever the modified square wave is outside the sine wave, the extra power can't be used by the motor and produces heat instead of torque. When the modified square wave is inside the sine wave, lower torque is produced resulting in a loss of power. The final result is that the spindle motor runs hotter than normal and produces less power than normal.

When manufacturers first began using high frequency spindle motors, they discovered that these motors did not seem to have as much power as standard synchronous motor driven spindles. A five horsepower synchronous motor powered spindle seemed to have much more power than a five horsepower high-frequency spindle. How can this be if both motors are the same horsepower?

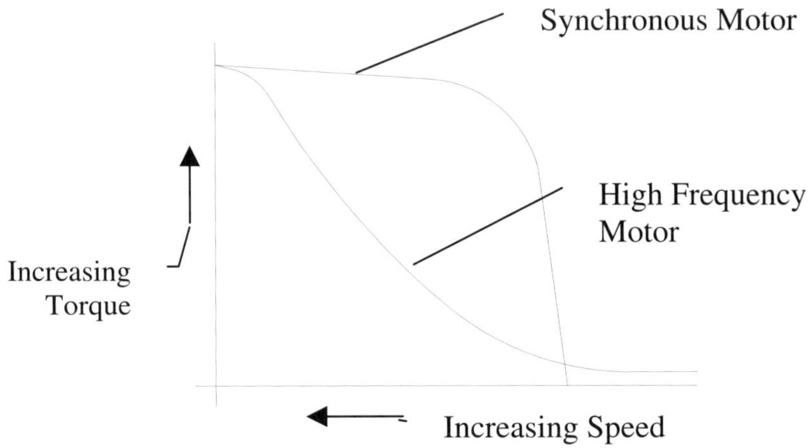

Torque versus Speed Curves

Horsepower is essentially torque times speed. It is determined by using the rotating torque available at full speed. In most cases this is a purely theoretical number determined using the amount of power the motor draws. As you load the motor by cutting, for example, the speed is reduced. The amount of torque available from the motor at these reduced speeds also changes. A graph of the available torque at various operating speeds is called the torque-speed curve.

The torque-speed graph of a synchronous motor shows that as speed is reduced, torque remains relatively constant until a certain point where the torque drops off rather quickly. On a high-frequency motor, however, torque drops off as speed is reduced. Since horsepower is speed times torque, a synchronous motor's horsepower drops off as it begins to cut because the rotation speed is lower. The horsepower of a high-frequency motor, however, drops off much faster since both components of horsepower are lower. Not only the rotation speed but also the torque is lower.

Thus, both motors have the same horsepower rating but when actually cutting there is less horsepower available from a typical high-frequency motor than from a synchronous motor. Thus, it is common to see ten or fifteen horsepower in a high frequency spindle when a five horsepower manual router operated by a synchronous motor has more than enough power for most applications.

How much actual power is required for CNC routing? In measurements of current draw we found that the actual power consumed by a properly operating system is much less than expected. Most applications required between one and two horsepower with the most demanding applications seldom using more than five horsepower. From a practical standpoint this means that most router spindles offered on machines today are capable of handling the demands of normal CNC routing. Spindle life, on the other hand, should be determined from actual field experience by current machine users. It is virtually impossible to determine from design parameters alone whether or not a particular spindle design will provide an acceptable life.

And now we will address the final area of a CNC router, the CNC control. This is likely the most technical and most complex of all of the technologies involved with this type of machine. In addition, this area is changing rapidly so technical features that are not even imagined today may very well be common in a year or two.

The very first machine controls were not CNC but were NC. NC stands for Numerical Controlled or controlled by numbers. These controls were electronic machine controls that responded to numbers that were input. The original machines read these numbers from punched paper tape. Holes punched in the paper tape corresponded to numbers. These were read by the control and then executed.

The early NC machines had no logic or control functions in the control at all. All information and techniques for operating the machine were carried by the punched tape program. This included functions such as acceleration and deceleration. Early machines tried to operate at the programmed feed speed instantly with no provision for controlled acceleration and deceleration. To prevent damage to the machine from these instant velocity changes, the programmer was forced to program short line segments at increasing speeds until final feed speed was achieved and then program additional short segments at the end of the motion each at slower speeds until the machine was stopped. The total length of all line segments then comprised the length of movement of the axis. As you can see, programming was an involved process with much opportunity for error. These early controls could seldom control more than two axes.

Once a computer was added to the NC control it became known as a CNC control. CNC stands for Computer Numerical Control indicating that the control is equipped with its own computer. Although some very expensive systems based on minicomputers were being experimented with in the metalworking industry, the very first CNC control based on the newly invented microprocessor was introduced by Thermwood in 1976. This system used an Intel 8080 microprocessor and allowed the control to calculate the acceleration and deceleration thereby making programming much easier.

Today, all decent CNC controls automatically handle acceleration and deceleration. In the past, considerations as simple as this were a major programming hurdle. This simple capability is so fundamentally important that I would strongly advise against any control that does not provide this most basic capability.

As computer technology advanced, CNC capability increased. Each control manufacturer, however, developed the technology along their own proprietary lines. By the time the PC started to develop and PC standards were established, the machine control people had already established a different approach. The approach established for CNC controls was primarily intended to obtain strong real-time performance.

Real-time in the computer world means that not only does a process need to occur but it needs to occur within a specified period of time to function correctly. A PC type computer product is only concerned with performing a task. It is not concerned with performing the task within a specific period of time so that it can control something like a moving machine. Once a machine is moving, the control functions must occur fast enough to keep up with the movement of the machine. Whenever a computing process must coordinate with a physical process, it is known as a real-time application.

CNC controls were designed primarily for real-time computing. Early PCs were too slow and their operating software too inefficient to allow their use in the real-time machine control applications. Machine control was left for specialized proprietary CNC controls.

As PCs became more popular, many low cost peripherals such as hard disk data storage became available. These devices also held advantages for CNC controls, however, since the controls were not built using the PC standards, the devices could not be simply added to the control. For the control manufacturer to use the devices on their own products, they needed to develop the most fundamental interface and driver technologies for each individual device. The cost of this effort was high and the relative volume of CNC controls compared to PCs was quite low. As a result, the cost to the customer for a device, a hard disk for example, was ten to twenty times as much as the same device for a PC. Customers had a hard time understanding this.

At the same time, the number of different devices coming to market was increasing dramatically. It was all but impossible for control manufacturers to integrate but a tiny fraction of the available technology. CNC controls seemed to be falling behind the fast advance of the personal computer industry.

Eventually the processing speed of PCs became fast enough that even with the poor real-time response of their operating systems, they were able to provide basic CNC control. At this point the CNC performance was substantially less than the existing proprietary controls, however, they were now able to easily integrate peripheral devices. The machine customer was now faced with a decision. Should I buy a proprietary control that is incompatible with most of today's PC products but provides good machine performance, or should I buy a PC based control that is compatible with much of today's low cost PC technology but offers a lower level of machine control.

This is where things get a little more complicated.

The manufacturers of proprietary controls quickly realized the threat that the new PC based controls held, so they developed an approach to try and achieve some of the compatibility of the PC based controls. Instead of basing their controls on a PC they instead added a separate PC to the system and then connected that PC to their control through a serial line. The way this works gets a little complex, however, it does offer some of the advantages of the PC based control to the proprietary controls. It also creates some bottlenecks that can negatively affect machine performance.

To understand these bottlenecks we must understand the fundamentals of how a CNC control processes part programs. A typical control structure maintains the part program in main memory. This memory offers very fast data transfer to and from the central processing unit or CPU. To execute a line of NC code, the line or block of code is read from memory, interpreted by the CPU and then converted to control instructions that are sent to the axis drives. Processing this block of data takes a certain amount of time. The time required to process a block of NC code is called block processing time.

In the past, block processing time was of little concern. A typical program was comprised of a few hundred lines of code, each representing relatively lengthy motions. With the increased use of CAD/CAM systems and the use of CNC routers for surfacing and carving, this is no longer the case.

The way that a CAD/CAM system creates a curved machine motion is to create a series of very short straight lines all connected together. These lines may only be a few thousandths of an inch long and there may be thousands upon thousands of these segments in a single curve. Executing this program now begins to create several problems.

The first problem that occurs is that the size of the program becomes very large. The overall program size is determined by the amount of CAD/CAM generated curved motion and the length of the straight line segments. Obviously the more curved motion the larger the program. Also, the shorter the line segment the more segments required per inch of motion and the larger the program. Shorter line segments generally mean smoother motion but at the cost of larger programs. Increasing the length of the line segments can reduce program size, but the result may be a rougher program.

Large programs create problems for many CNC controls. The first problem occurs when the program will no longer fit into main memory. Some method must be found for either breaking the program up into two or more smaller programs that will fit into available memory or finding some system for downloading the program into machine memory a segment at a time while the machine is executing. The most common way of doing this is to attach a PC to the control through a serial link. The large program is stored on the PC and then fed a segment at a time to the main control memory once the existing code in memory has been executed. Before the PC there were computer systems designed to store large programs and “drip feed” them to the main processor but these were used almost exclusively by the aerospace industry.

Although this system works it creates some performance problems. With very short line segments and high feed speeds, data is executed at a high rate. Many times this rate is faster than the ability of the serial link to transfer replacement data into memory. It is possible to run out of data to execute and the machine must stop and wait for new program data to load before continuing execution. These pauses in motion can cause quality problems when working in many materials. They also slow program execution since the performance of the control determines cycle time rather than the cutting process.

Another approach that works well is the approach that Thermwood has taken in its SuperControl. They store the program on a very large hard disk that is part of the main control and then operate from the hard disk as if it were main memory. By eliminating the data transfers common with the other approach, virtually any size program can be executed as if it were located in main memory. Although hard disk access is slower that RAM memory access, it is much faster than necessary to keep up with almost any block processing speed. Large complex programs no longer slow the control and limit production rates.

Another technical feature that becomes important in these circumstances is the speed at which the control can process a block of NC code. This is simply a measure of how fast the control can retrieve and execute a single line of NC code. It is called block processing speed and it determines the maximum feed speed of the machine executing a specific program. If block processing speed is ten milliseconds this means that the control is capable of executing 100 lines of code per second. If each line represents a segment .010 of an inch long, the maximum feed speed is one inch per second or 60 inches a minute. This is a relatively slow feed especially for a complex part such as a three-dimensional wood pattern.

In general, you need to use the longest line segment that will yield an acceptable surface finish and the fastest control you can get. Before the advent of CAD/CAM generated programs, block processing speed was not particularly important. In three-dimensional metalworking, feed speeds were slow enough that the control processing limitation never really came into play. Only when we began to try and execute CAD type programs at high speed did the limitation begin to impact performance. This is another area where the demands of CNC routers far exceed the requirements for metalworking.

Thermwood's SuperControl points the way to the future of CNC controls. The architecture is not a PC based control nor is it a proprietary control. Instead it is a hybrid, combining the advantages of both systems into a single package.

This system uses a PC front end just like the metalworking controls, however, the PC is connected to the main control not through a link but by existing on the same internal buss as the main control. This make interchange of data almost instantaneous and causes the PC front end and the proprietary CNC control to operate as a single unit.

This architecture offers the compatibility of a PC with the raw processing power of a high-end proprietary CNC control. The control is capable of handling massive programs hundreds of megabytes in size, and it is capable of executing these at extremely high block processing speeds. This system also offer two features that have long been the goal of the control industry, multi-tasking and open-architecture.

These two features open a new world to CNC control and offer the potential for revolutionizing the industry once the capabilities are fully understood.

Multi-tasking simply means that the control can execute more than one independent program at the same time. Open-architecture means that a program that was not written by the control manufacturer can be executed on the control. Although both of these seem simple enough, the impact that they have on both capability and productivity can be substantial.

We take open-architecture for granted since it is the basis for personal computers. Every PC is inherently designed to run a variety of programs. It is difficult to understand that a CNC control is not like that at all. CNC controls are dedicated machines intended to run only the software written by the control manufacturer and no other. The control computer tends to be specialized making it perform very well, but only in a rigid carefully controlled environment. Creating a CNC control that can also run other software is very unique today.

Multi-tasking has become somewhat commonplace with the advent of Windows NT and Windows 95 on desktop computers. Thermwood selected OS-2 as the operating system for its SuperControl because it has a much more predictable and more robust multi-tasking operation than any other commercially available operating system today. CNC controls operate in a real time environment and the ability to properly allocate processor time is absolutely essential. Operating systems are constantly improving and I believe that several systems capable of multi-tasking in a real time environment such as CNC controls will become available in the near future.

On the surface, multi-tasking allows you to call up and edit one program while the machine is executing another. Real multi-tasking, however, holds promise well beyond this limited view.

An electronic operator's manual could be called up and procedures reviewed by the operator while the machine is in operation. The control could be placed on a network and messages sent and received by the operator while the machine is in production. A small video camera could monitor the cutting process inside the dust hood and display the resulting image directly on the control screen while the control is in production. The control could be queried in the background and complete data about the performance and production status could be uploaded across a network without the machine operator needing to assist the process. A separate program could keep tract of the actual movement of each part of the machine and alert the operator to the need for lubrication or other routine maintenance only when it was absolutely necessary. A tool management program could keep track of the actual cutting time on every tool bit and alert the operator when the tool life has been exceeded. A CAD/CAM or other code generating program could be run on the control creating the next program while the current program is executing.

All of these capabilities are not only possible with multi-tasking but are actually being used in production on the Thermwood SuperControl. Most control manufacturers are currently working on their own hybrid controls with multi-tasking and open-architecture. Within a few years these capabilities will be commonplace and even more new innovations will be added to the list.

Another advantage of the PC-based control or a hybrid control with a PC front end is that the part programs can be assigned alphanumeric names and can be stored in a directory structure. Most proprietary controls limit part programs to numeric names, which are not particularly easy to identify. When both letters and numbers can be used to name a program, it is much easier to locate the program you want.

The directory structure allows programs to be grouped in specific directories by design, type of part or finished product. With the advent of mass program storage, it is possible that a factory can generate and store in the control thousands of different programs. Locating and identifying the program you want can become difficult and time consuming if you do not have a system that makes this easy.

An open architecture control should allow the installation of a bar code reader directly on the control. A bar code containing the key strokes necessary to load the correct program can be printed on top of the work order. With this arrangement, all the operator must do to locate and load the correct program is to scan the bar code. This process takes a few seconds and virtually eliminates program loading errors.

Another capability that is showing up on newer controls is the ability to execute parametric programs. Parametric programs are part programs where all required dimensions are not part of the original program. For example, instead of writing a program for a kitchen cabinet door that specifies the width and height in the program, we write the program using the word "width" and "height" in place of the actual numbers. These words are called variables. All dimensions are defined using the variables. For example the center of the door can be called "width divided by 2". As you can see this will give you half the width on any size door.

Once the program is completed using variables, you can assign actual numbers to "width" and "height" and machine a specific size door. You can then assign another set of numbers to the variables and use the exact same program to machine a different size door.

Kitchen cabinet doors are a good example of the power of parametric programming. A typical kitchen is made up of dozens of different sized cabinets each requiring a different size door. Using normal programming techniques, each different size door will require a new program. A particular door style may require over a hundred different programs to handle all the different sizes required. With parametric programming, a single program can not only manufacture every door needed but also any size in between.

Although this example is for a two-dimensional part, the same technology can be applied to three-dimensional parts.

In the past, parametric programming was used in a stand-alone parametric CAD program, quickly creating all the programs needed for a set of cabinets. The problem is that you must first create the final execution program and then load that program in the machine control to cut the door.

To address this problem, Thermwood in its SuperControl has created a programming language called the Advanced Function Language. The Advanced Function Language or AFL is actually a full function computer programming language that can be incorporated directly into the NC part program. Using this program, variables can be defined, values assigned and mathematical functions performed. Parametric programs can be written using AFL and executed right on the control.

The Advanced Function Language offers much broader power than simple parametric programming. For example, not only can it define a variable but it can open a window and request that the operator input the width of the door which it then assigns to the program to cut the part. For an even more automatic operation, AFL could use a sensor to detect the edge of the door. Move an axis until the sensor detects the edge, use the position of the axis to determine the width of the door and automatically assign the dimension to the variable. With this system working, all you need to run any size door is to load the blank and press the start button. The AFL program measures the size of the door, loads the proper variables and cuts the finished product.

Using the Advanced Function Language, every part of the machine can be queried and controlled. Basic machine functions can be added for specialized applications. Truly, if you can imagine a function you wish to perform, you can very likely add it to the control using the Advanced Function Program.

Another feature of AFL is the ability to create macros. Macros are actual machine programs that can be assigned unused “M” or “G” code names and then they can be called by other programs. The real power of macros is very subtle but they are very powerful.

To understand the value of machine and tooling macros you must first understand how differences in the actual physical construction of CNC machines is handled by programmers. No two machines are exactly the same, however, a machine with a single head will require very little effort to run the same programs on different machines. When a second or third head is added, problems become greater. It is virtually impossible to manufacture two or more machines where the spacing between heads is exactly the same. When a program switches from using head one to using head two, it must shift the heads to allow for the distance between them. Normally this shift is a motion within the program.

This program with the shifts between heads programmed in will only work on one machine. A second machine will require slightly different shifts between heads so it will require slightly different programs. This means that a program must be generated for every part for every machine. Five different parts that can run on five different machines will require twenty-five separate programs. This is getting quite complex.

One way of simplifying the programming effort is by using a CAD/CAM system. The Computer Aided Design or CAD portion of the system creates the actual part geometry. The Computer Aided Manufacturing portion of the system determines the tool path and tools required to manufacture the part. A final program called a Post Processor inserts the shifts between tools for a particular machine. A Post Processor program must exist for every machine. It contains the shift distances for that particular machine.

Now, a program can be developed all the way through the CAM system and then simply run through a different Post Processor for each machine it will run on. This simplifies program development but still results in twenty-five different programs that must be tracked and controlled so that the program is not run on the wrong machine.

Should a machine crash or a head change result in slightly different shift distances for a machine, the Post Processor must be modified to reflect the new distances and then all old programs must be re-posted using the new post processor. This can be a time consuming task when a company has thousands of existing programs. Again, accurately tracking the status of a program is difficult. An error will result in parts that look okay but are slightly off dimensionally.

Tooling macros address this problem in a different way. Instead of imbedding the shift distances in the actual part program, the macro is actually a short program that performs the shift. These programs are left in the machine but are called by the part program. For example, you may have a tooling macro that shifts to tool two. This macro will have a specific name. The macro that shifts to tool two will have the same name on every machine even though the actual macro itself is not the same. Each macro will perform the steps necessary to shift to tool two on that specific machine.

The part program simply calls the appropriate macro to shift to tool two. It will perform the proper steps regardless of which machine the program is run on. Thus, the exact same program will work on every machine even though the machines may be physically different. An extreme example of this would be two machines, one with the second tool mounted to the right of the main head and another with the second tool mounted to the left of the main head. Without tooling macros, a totally different program will be needed for each machine. With tooling macros, the macro called by the program will shift the head to the left on machine one and to the right on machine two. The same program will work on both machines even though they are quite different.

In the case where a tool is remounted in a different location, it is only necessary to modify the tooling macro for all programs to operate properly. It is no longer necessary to re-post all existing programs.

Another major benefit of tooling macros is that only one program exists for each part. That program will result in an accurate part regardless of which machine it is run on. The possibility of running a program on the wrong machine is totally eliminated.

This is a highly simplified version of how macros actually work in modern controls. The macro not only considers the tool it is going to but also the tool that is currently being used. The shift direction and distance will be different moving from tool one to tool two then if moving from tool four to tool two. Macros are also used to control spoilboard thickness, fixture location and a variety of input-output functions.

A control capable of handling macros can substantially simplify both programming and operation and will make errors much less frequent.

Thermwood's Advanced Function Language can go one step farther than a macro. Any set of program instructions can be given an unused "M" or "G" code name and essentially turned into a custom macro for a specific application.

There is another feature that is currently available from the SuperControl and may soon be available from other control manufacturers. This capability is the ability to execute five-axis matrix transforms in the control.

To try and understand this, let us use a simple example, tool length compensation. Let us assume that we would like to put different length tools in the router and then have the control adjust the program for these different lengths. If you think about it, if the tool is pointed straight down, all you need to do is to adjust the vertical "Z" axis for the new length. You simply add or subtract the difference between the current length of the tool and the tool length used when the part was programmed.

If the tool is pointed at some angle in space, however, this process becomes much more complex. We must now adjust all three linear axes so that the tool moves precisely in the direction it is pointed just enough to compensate for the new length. If the cut path involves one or both rotary axes, these calculations are constantly changing.

Calculating the amount of adjustment required for this involves five by five matrix math. These calculations are quite involved and require substantial computer time even using a math co-processor, which is built into most new microprocessors. The time required to perform these calculations generally slows block processing time enough to make it impractical.

To achieve real-time matrix transforms, Thermwood added a custom matrix math co-processor chip to its control. This chip constantly performs matrix transforms on the current position. When the main processor needs a transform, it simply reads it from the custom chip. This approach allows matrix math without slowing the execution speed of the control.

For this particular example, it is possible to simply enter the new tool length and then recalculate the entire program ahead of time. This is called "pre-processing". This is accomplished before the machine is run so the amount of time it takes is not significant. Then it is simply necessary to execute the modified program. No loss of speed or block processing time occurs.

In the future, however, applications will develop where performing axis transforms real-time during program execution will become necessary. For example, if you are trying to accurately trim a three-dimensional part, remove a mold line for example, but the actual part varies in size. You can use some type of sensor to measure the part but you must then transform the program as you are executing it to match, not the size that was programmed but the actual size of the part you are cutting.

Again, you could measure the part by executing the program using the sensor and then transform the program and trim the part. This, however, at least doubles the cycle time and is unnecessary if real-time transforms are possible.

Here we are looking into the future somewhat, however, as more capabilities become available the CNC router will be able to address more and more applications. Advancing control technology will be a key component to three-dimensional trimming and machining in the future. The production techniques that are evolving will rely heavily on advanced features of new age CNC controls. As computer technology continues to advance at an ever increasing speed, we can look forward to even more capable controls and even more possibilities for improved production systems.

Chapter 3
Holding the Part

Holding the Part

In order to machine a part on a CNC router the part must be held securely in place. This seems obvious, however, this is the one area that most new CNC router users underestimate. We are combining all of the various methods of holding parts for machining under the category "fixturing". The hold down system has a significant impact on part accuracy, quality of finish and, at times, even feed speeds and tooling life. There are two things that should be understood about part hold down; holding the part securely is important and; there is no one system that will properly hold all parts.

There are two fundamentally different types of parts that must be held in place. The first is the flat part or a sheet of material and the second is a three-dimensional or molded part. The fixturing systems for each are similar, however, the molded three dimensional part normally requires somewhat more complex arrangements.

There are a wide variety of hold-down systems available today. Each works well for a certain type of part. We will examine each of these systems and explain which system works best with each type of part.

The purpose of the hold-down system is to hold the part in place for machining. This is so basic that people underestimate what holding the part in place means. A vacuum hold-down where the part rests on rubber seals may allow the part to move or wiggle slightly on the soft seals. This can easily result in excessive tooling marks, chatter and a poor quality edge. When this occurs, people tend to blame the machine, the cutters or the material where the hold down system is the real culprit. It is also possible that under the pressure of cutting, the part may move slightly resulting in a loss of accuracy.

Holding the part rigidly in place is important. It is much more important if you are trying to achieve a smooth machined edge and less important if you are simply looking for a specific shape for a non-cosmetic part.

Some materials require higher cutting forces than other materials. These materials will require a more rigid hold down system. Some materials will vibrate or chatter when cut. The hold down system will need to support the part and eliminate the chatter.

When machining flat material, generally the CNC router table is covered with a sheet of plywood, particleboard or other material on to which the parts are clamped or the fixtures clamped. This table covering is referred to as a spoilboard or a wasteboard depending on the machine manufacturer you are talking to. We will call it a spoilboard in the book.

The major purpose of the spoilboard is to protect the table when you are cutting all the way through a part. Many times the hold down method is incorporated directly into the spoilboard. Other times a fixture for the part is mounted to the spoilboard. There are other times where the spoilboard is an integral part of the hold-down system.

Fixtures for three-dimensional molded parts are normally bolted directly to the machine table.

The most basic hold-down system is to simply nail or bolt the part down. Some parts have interior holes or are not cosmetic at all and it is possible to simply attach the part to the spoilboard mechanically. Except for machining very small parts, this is seldom used for production but is common for one-up parts or prototypes where the operator doesn't want to stop and build a fixture for one part. Obviously, the time required to fasten the part down and then remove it make this system impractical for most production operations, not to mention the hole it leaves in the part. There are exceptions to this however. I have seen small Derringer pistol handles held down through a counterbored hole in the center. These were very small parts that could be completely machined using this hold-down method.

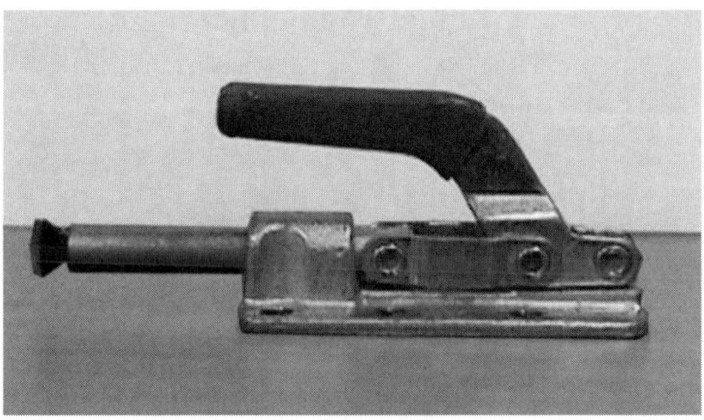

Side Activated Hand Clamp

Hand operated clamps offer a practical method of mechanically holding the part in place for a much wider variety of parts. Properly applied, these offer a highly rigid support for the part. There are a wide variety of mechanical clamps available however they tend to fall into two basic types, those that clamp the part to the table top or to a holding fixture and those that push the part into a fence or guide.

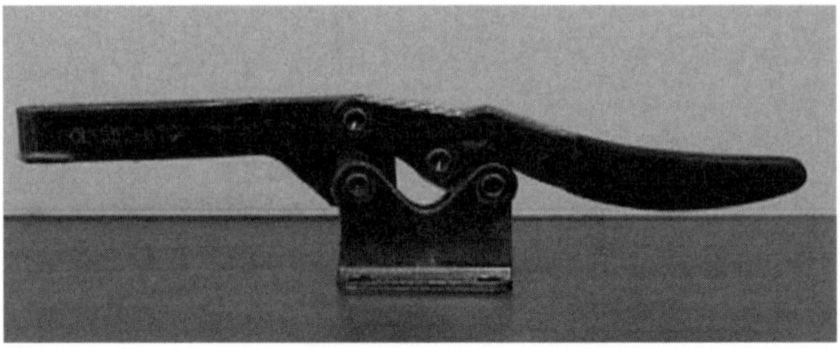

Top Acting Hand Clamp

Generally, if you are going to work on the edge of the part, you will need to clamp the face of the part to the table top leaving the edge exposed. If you are going to work on the face of the part you will need to clamp the edge of the part into a guide or fence. In this arrangement, the guide or fence should be designed to pull the part down to the table top as well as clamp the edge.

On a molded part, many times it is possible to clamp the scrap or trim material that will be trimmed away instead of clamping the actual part. The only additional requirement in this arrangement is to provide some means of holding the part while the final cut is made. The final cut is when the part is actually cut free from the trim material. If you don't hold the part at that point, it can move leaving a pointed tab on the part.

If you are using a guide fence, it is a good idea to leave the corner open to prevent chips or sawdust from collecting and affecting the positioning of the part. In the same manner, it is a good idea to relieve the bottom edge of the fence where it meets the tabletop to minimize the effect of the small amount of chips or sawdust that will collect in these areas.

Mechanical clamps offer a good alternative if you are going to work on one, or perhaps two edges of a part or if there is scrap material that can be clamped around the part. If, however, you need to machine all the way around a part and there are no good clamping areas, mechanical clamps don't work well since on at least one edge the clamp itself will be in the way.

Air, or pneumatic powered clamps have been used in these circumstances. By placing one clamp along one edge of the part and a second clamp along a second edge, the entire perimeter can be machined. As the cutter approaches the first clamp, it retracts under machine control while the second clamp holds the part in place. Once the edge has been completed, the first clamp again clamps the part down.

The process is repeated with the second clamp. As the head approaches, the clamp retracts while the first clamp holds the part. This does work, but there are simpler methods of holding a part that must be machined all the way around.

The advantages of mechanical clamps are that they are inexpensive, easy to understand and offer a highly rigid mounting when properly applied. They may be the only practical hold-down method for small or odd-shaped parts that do not have enough surface area for some of the vacuum hold-down systems.

The most common system for holding parts on a CNC router is conventional vacuum. Vacuum does not actually hold the part. Vacuum is simply the absence of air. The approximately 20 mile thick layer of air surrounding the Earth has weight. At the surface of the Earth it weighs about 14 pounds per square inch. The air, being a fluid, pushes on everything in all directions equally so no force is exerted on objects.

When the air is removed from one side of an object, the air on the other side now pushes against the object. This is the basis of vacuum hold-down. The part to be machined is sealed against the tabletop or a fixture and then the air inside the seal is removed using a vacuum pump. The air on the outside then pushes the part against the fixture.

Essentially with conventional vacuum all you are doing is weighing down the part using the air pressure of the atmosphere. How much weight does the air pressure actually put on the part?

As we said earlier, the air weighs about 14 pounds per square inch. This level actually varies each day and can be determined by noting the barometric pressure in the weather report. Conventional vacuum actually works slightly better some days than others. Thus if you could achieve a perfect vacuum you could achieve about 14 pounds of force per square inch. It is virtually impossible to achieve this level of vacuum however. In practice with a well-sealed system, a good pump and material that does not leak can consistently achieve about 10 pounds of force per square inch.

Thus a one foot by one foot part has 144 square inches of area if the seal is located at the very edge of the part. At ten pounds per square inch this part would be held in place with 1,440 pounds of force. As you can see, this is a considerable force. It is like piling over 1,400 pounds of weight on the part and yet the part is left completely accessible for machining, free of weight or clamps.

Obviously as the area inside the seal gets smaller, the amount of holding force gets less. A six-inch by six-inch part has 36 square inches inside the seal and so is held in place by 360 pounds. Heavy cuts in hard material may produce enough side force to move this part. A three-inch by three-inch part has nine square inches or ninety pounds of holding force. The hold down method is now getting quite marginal. There are a few techniques that help to hold smaller parts.

First, if you are machining a three-dimensional part on a three-dimensional fixture, the shape of the part may help hold it in place. It takes substantially less force to hold a molded part to a form fitting fixture than it takes to hold a flat part on a flat tabletop.

When machining a flat part, a common method to assist vacuum hold-down when the back surface of the part is not critical is to install barbs or metal points in the fixture so that they stab the back of the part when it is loaded. The air pressure pushes the part against the barbs making it more difficult to slide. Forces during cutting are generally side forces trying to slide the part. The barbs can help prevent this.

If the back surface is more critical, adhesive backed sandpaper can be installed on the fixture inside the seal. Again, the air pressure presses the part against the sandpaper, increasing friction and making it more difficult to slide the part.

Eventually part size gets too small and it is necessary to resort to another hold-down method. There are also other factors that reduce the effectiveness of conventional vacuum and must be considered.

As we have seen, to make conventional vacuum work we must remove the air on one side of the part. This may be much more difficult than it sounds. Air, being a thin fluid, is trying desperately to get back into the evacuated area. This is one reason for the seal. Air leakage through the seal area can reduce the effectiveness of vacuum hold-down to the point where it no longer functions effectively. The seal material must be soft and pliable enough to prevent air leakage and yet strong enough to withstand repeated loading and unloading of parts without becoming damaged.

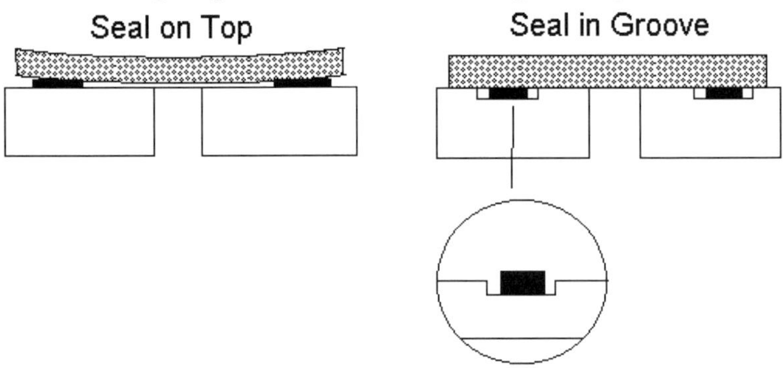

The seal material should be placed in a machined groove cut into the fixture rather than directly on top of the fixture. No more than one third to one half of the seal height should extend above the top surface of the fixture. When the part is pulled down the seal will then collapse to perhaps one half of its normal height. If the seal is placed on top of the fixture, directly between the fixture and the part, when the part is pulled down, thousands of pounds of force will be pressing on the seal trying to smash it completely flat. This will quickly break down almost any seal material. It will also cause the part to warp with the center pulled down to the face of the fixture but the edges raised up on the seal. Neither of these is a good thing.

The seal area, however, is not the only source for air leaks. The part itself can be a problem. Many wood and composite wood materials are somewhat porous. Air can actually leak through the part. At this point the flow capacity of the vacuum pump determines whether or not the system will work. Some level of vacuum can be established. Whether or not this level will be adequate depends on many factors including part size, pump capacity and amount of leakage. At times, people have laid brown paper or newspaper on top of the part to help seal its front surface and reduce air leakage.

Air can also leak through the fixture. Many companies like to make vacuum fixtures from high quality plywood. The hole through the center of the fixture through which air is removed exposes the edge of the plywood to vacuum. It is common for air to leak between the layers of plywood rendering the fixture ineffective. In these cases it is necessary to seal the vacuum hole and the top surface of the fixture in order for it to work properly. The hole and plywood edges can be sealed with simple wood filler and the surface can be sealed with any good paint sealer.

Conventional vacuum offers a good, reliable, low cost and effective hold-down method for CNC routers. The only real disadvantage to conventional vacuum as we have described it so far is that you are required to build a vacuum fixture for each part you wish to process. We will discuss some methods of building fixtures for molded three-dimensional parts shortly. For flat parts or material, several methods have been developed to try and get around the requirement of building a fixture for each part.

Pod systems use vacuum pods of approximately three-inch diameter. These pods are mounted on some type of adjustable slide mechanism. To fixture a part, pods are moved and secured under the part. Pods may be round, square or rectangular depending on the manufacturer. Pods work reasonably well for larger parts although setting up a pod system for several parts may actually take longer than loading a single vacuum fixture that has already been built to accommodate those parts. If parts are to be manufactured more than one time, building a fixture may be more economical. If parts are only going to be run once or twice, the pod system may be the better choice. You should also note that the amount of holding force with a pod system tends to be less than a custom-made vacuum fixture. The pods are also connected mechanically to the machine, so any movement at the attachment point may result in a less rigid hold-down.

Carter flop-pod System

A patented system called the Carter Flip-Pod System offers a novel approach. The Carter Pod consists of a molded plastic pod approximately three inches in diameter. This pod has a lip along one edge that fits into a recess machined into a vacuum plenum that covers the entire tabletop. The pod can be inserted into the plenum in two different orientations. In one orientation, the pod fits down into the plenum with its top surface flush with the tabletop. A small rubber ball inside the pod covers the vacuum access hole sealing it off. In this orientation the pod is inactive.

If the pod is flipped over, however, it extends a couple of inches above the tabletop exposing a vacuum seal. The ball inside the pod falls from the vacuum hole, opening it up and the pod is active. Setting up this system requires that the pods you wish to use be simply flipped over.

Again, the system works reasonably well and because the pod is machined into the tabletop, it offers a fairly rigid mounting. Just like the other pods, however, the amount of holding area is less than a full custom vacuum fixture.

Both a custom vacuum fixture and a pod system share the requirement that they be set-up each time the part to be machined changes. When the production batch size becomes very small, one or two parts, this can become a major disruption. A system that is more universal would be a major benefit.

Earlier we mentioned that air would actually leak through certain materials. This is especially true of certain types of particleboard and MDF. This fact can be used to develop a type of universal hold-down although it has some limitations.

Turbine Vacuum Pump

To make this system work, a different type of vacuum pump is needed. The conventional vacuum pump creates a fairly high level of vacuum but it does so with relatively small flows. A typical conventional vacuum pump has flow levels of 30 to 50 cubic feet per minute. To make this universal vacuum system work, flows of 500 to 1,000 cubic feet per minute are necessary. At these flow levels, the actual amount of vacuum will be substantially less than is normal with a conventional vacuum system.

Universal Vacuum Plenum

A particleboard tabletop is mounted to a vacuum plenum. The vacuum plenum is a sheet of plastic material about ¾ inch thick into which has been machined a pattern that looks much like a waffle iron. This exposes much of the underside of the particleboard to the vacuum chamber while still supporting the material. When the universal vacuum pump is connected to the plenum and turned on, air is constantly pulled through the particleboard. The flow capacity of the pump is actually greater than the total leakage across the entire tabletop. This maintains a vacuum under the tabletop and creates a very thin low-pressure area on top of the table. If a part is laid in this low-pressure area, the amount of air on the underside is less than the amount on top of the part and it is held in place. This works without the need for seals. It also works even if there are grooves cut in the particleboard top from previous parts.

The amount of holding force available from universal vacuum is substantially less than that available from conventional vacuum. Whereas conventional vacuum can develop up to ten pounds per square inch, universal vacuum generally works between two and four pounds per square inch. A two-foot by two-foot part is held in place by one to two thousand pounds of force, which is more than adequate for most applications. A one-foot by one-foot part is held in place by three to five hundred pounds of force, which is nearing the lower limit of hold-down requirements.

Although universal vacuum can hold a part without need for fixtures or set-up, its lower size limit is less than conventional vacuum. Also, the part must be relatively smooth and flat.

The low-pressure area developed on the tabletop by universal vacuum is only a few thousandths of an inch thick. For a part to be held in place it must move into this thin low-pressure area. Parts with a rough surface, such as plywood, cannot move close enough to the tabletop and thus will not be held in place properly. Also, a part that is warped will not move into the low-pressure area and thus will not be held down properly. In general universal vacuum works with smooth, flat parts that are at least a square foot in size. Once these requirements are met, the system works very well in practice.

Some manufacturers that do not offer universal vacuum have clamed in print that the particleboard top used in universal vacuum tends to fill with sawdust over time and quits working. In actual practice this does not seem to occur. The amount of force created by the vacuum and the airflow at any point on the surface are both quite low. Because of this there is not enough force to draw the sawdust that falls on the tabletop deep into the pores of the top. Simply blowing on the surface with compressed air seems to easily clean the surface and problems with plugging the particleboard top have simply not developed in practice.

If parts change often and are both large enough and flat, universal vacuum offers an extremely simple and easy hold-down system without the need for set-up or fixtures.

The final hold-down system for flat material that we will discuss is also somewhat universal in nature. It was originally designed to hold down plywood boat bulkheads for the marine industry. These parts are machined from large sheets of plywood, are relatively large and change almost every cycle. The plywood used was almost universally warped and so a method was needed that would press the parts down to the tabletop for machining.

To accomplish this, a set of four rubber-coated rollers was placed across the width of the worktable. Two of these rollers were placed in front of the router spindle and two of the rollers were placed in back of the spindle. The router spindle could move back and forth between the rollers and when the spindle moved front to back the rollers moved with the spindle. The rollers actually pinched the part to the tabletop rolling over the top of the part as it moved front to back.

Roller Hold Down without Covers

Although the system works well it also has some limitations. In order to hold a part properly, at least two rolls must contact the part. Because of clearance requirements this means that the part must be at least ten to twelve inches long or is will simply slip between the rolls. Even with parts that are of proper length it is not uncommon for the parts to move slightly when they are cut loose from the sheet. There are some programming techniques that can help should this occur. Parts should be oriented with their longest dimension perpendicular to the rolls. The cut on each part should be started in the center of the part where there is a roller on each side of the cut. Then, when the part is cut free, there is a roller holding each end of the part, which is more stable.

Another technique raises the cutter slightly about an inch before the part is cut free from the main sheet. This leaves a thin bridge between the part and the main sheet. This bridge prevents the part from moving during the final cut. In many cases, the machine can then reverse direction, lower and cut off this bridge. Since the bridge is very thin, little cutting force is developed and the chances of the part moving are reduced considerably. Parts too small to be properly held by the roller can be left attached to the main sheet by several of these bridges and then broken out by hand after the sheet is removed from the machine.

Very small parts can be machined using either vacuum or roller hold-down by not machining all the way through the part. By leaving a .010 to .015 inch web on the back of the part, the small parts can be machined easily. Then the sheet of parts is turned over and run through a wide belt sander to remove the web and separate the parts.

Early roller hold-down systems simply attached the rollers to the frame of the machine. The table bearings or the gantry bearings, depending on the machine design, resisted the clamping force of the rollers. As you increased clamping pressure to help the rolls hold the parts, you also increased the load on the axis bearings. This resulted in premature bearing failure in some machines. To address this issue, one manufacturer has installed a set of wheels that ride under each edge of the tabletop and pinch the table and parts between the rolls on top and the wheels underneath. In this arrangement the rolls add no load to the axis bearings regardless of the clamping force used.

Roller hold-down is commonly used when parts are to be cut from large sheets of material. In many cases, thinner material can be stacked and the entire stack cut at one time. Cutting irregular shaped parts from a full sheet of material can offer substantial yield improvements over cutting the sheet up on a panel system and then cutting the parts from the resulting blanks. When cut from a full sheet, parts can be nested to use the absolute maximum amount of material possible within the part itself.

There are several software packages that can assist in nesting a variety of parts based on production volume requirements. These nesting packages may either be able to run in the control, if the control is open architecture, or may be run on a separate computer and downloaded to the control. Each nest may be different and thus some type of universal hold-down system is needed. Since many times the material being cut cannot be held properly by universal vacuum, roller hold-down becomes the only alternative.

From an overall cost standpoint, mechanical clamps are the least cost method. These are followed by conventional vacuum and then universal vacuum. Roller hold-down tends to be the most expensive system. In the final analysis, however, the material and the part design may dictate the hold-down system rather than the system cost.

Fixturing and machining three-dimensional molded parts generally requires a form fitting fixture onto which the part can be mounted. Normally the molded part is held to the fixture using conventional vacuum, although some parts that do not have enough surface area can be clamped in place.

There are three different methods I have seen work in building these fixtures. The first is to actually machine the fixture from wood, aluminum or plastic material. This normally requires a CAD/CAM system to develop the necessary program. It can also be quite time consuming.

This method does, however, result in a very high quality, durable production fixture.

A machined holding fixture can have one subtle problem. If the fixture program is developed from original design data, it may not match the resulting molded part accurately enough to work properly. Several factors contribute to this problem.

First, a molded plastic part rarely ends up the exact size predicted by the CAD/CAM system. Variations in material, the molding process and the actual shrinkage compared to the theoretical shrinkage generally results a part that is close but not exactly the size intended. If we then machine a fixture for the theoretical size, it may not match.

On a thermoformed part, many times the mold supports the part on one surface and the trim fixture supports the part on the opposite surface. Variations in the molded thickness of the part, which are very difficult to predict, will also cause a mismatch between the part and the holding fixture.

Another potential problem is that it is quite difficult to machine an accurate fixture directly from theoretical

CAD/CAM data. We will cover this problem in much more detail in the chapter on accuracy.

One technique that can make this process practical is to first mold a part, then probe the part (using the CNC router) to capture the exact shape of the finished surface of the part. This data is then fed into the CAD/CAM system and the program for the fixture is developed.

If the fixture is machined on the same machine used for probing, any variations in the machine will tend to cancel out resulting in a relatively accurate holding fixture.

There are different methods of probing the part. The easiest is if the machine you are using supports a probe. If that is the case, it is simply necessary to move the probe around the part recording points. The probe can be used to actually move the axes of the machine so this process can be completed in anywhere from a few minutes to an hour.

If your machine does not support an electronic probe, a dial indicator can be mounted in the spindle and the part can be mechanically probed. To use this method, the machine is moved to a point over the part. Then the Z axis is lowered until the dial indicator achieves a reference point. Then, the positions of all three linear axes of the machine are recorded. This process is then repeated until a sufficient number of points have been recorded.

Obviously this second process is much more time consuming, but it does work.

Although this approach does result in a durable, accurate holding fixture, there are other, simpler methods of constructing holding fixtures.

The most common way of building a fixture is to cast it using the part itself to define the shape. Several materials are available for casting fixtures. Simple plaster has been used to cast fixtures. Normal plaster will shrink as it cures and is not generally suitable. The foundry industry offers different types of casting plaster including one that actually expands as it cures. These are used to allow for shrinkage of the cast metal.

A casting plaster that is used for fixtures should have zero shrink.

Plaster tends to be relatively inexpensive, however, the resulting fixture can become rather heavy. Plaster is also somewhat brittle making it susceptible to handling damage.

A common material used to make three-dimensional fixtures is a two-part urethane. Generally, the part itself is used to cast the fixture. If the part is not structurally strong enough to maintain its shape during the casting process, it is necessary to reinforce the outside of the part before casting the urethane on the inside. Also, curing urethane can generate substantial heat so even if the part is strong enough at room temperature, it may be necessary to reinforce it for the casting process.

A good way to reinforce the part is to smear a fiberglass reinforced gypsum over the outside of the part. This material is inexpensive and easy to use and seems to do a good job of reinforcing parts for casting.

After the fixture is cast, it will be necessary to cut both a groove for the vacuum seal and channels to help provide a path for vacuum to reach all areas of the part. One trick that works well is to stick an adhesive-backed rubber seal on the inside of the part along the seal area as well as along the vacuum channels. After the casting has cured, these seals can be removed resulting in channels in the fixture, which can be used for locating the vacuum seal and vacuum channels. This virtually eliminates the need to machine the fixture after casting.

A release agent must be applied to the inside of the part prior to casting to prevent the urethane from adhering to the part.

Generally, it is a good idea to keep the cast urethane cross section less than two inches. Thicker sections than this have a tendency to crack during the casting process. Filler blocks can be inserted into the casting to keep the cross sections to the proper thickness.

Some people have started to use castable silicone to make fixtures. This material comes in a variety of durometers from rigid to quite soft. The casting process is similar to cast urethane except the soft nature of silicone eliminates the need for separate vacuum seals. Also, conventional vacuum channels don't seem to work as well as urethane so a large number of holes are provided through the fixture to a vacuum chamber. The holes I have seen have been drilled, however, it might be possible to use straws or tubes attached to the inside of the part prior to casting to eliminate the need for drilling.

Another material that works well for three-dimensional fixtures is fiberglass. The process is similar to cast urethane in that the part is reinforced on the outside and vacuum seals and channels are attached to the inside. A release agent is applied to the inside of the part and then a gel coat is sprayed inside the part. Then layers of mat or chop are applied until an adequate thickness is achieved. Mounting blocks or brackets are then "glassed in" the fixture to provide a method of mounting the fixture to a base.

Each of the methods described require a mounting base to attach the fixture to the machine table. This base should be aluminum or steel. Although we have made many attempts to use wood bases, including high quality pattern plywood, changes in temperature and humidity have resulted in problems keeping the base straight and flat.

If the fixture is to be used on more than one machine, it is a good idea to provide setup reference points on the fixture. Setup reference points consist of three different reference points located at three corners of the fixture. A setup program is written that moves a tool or a pointer located in the spindle to just touch each of the three points.

When it is time to either remount the fixture or move the fixture to a new machine, the setup program is loaded and the head moved to the first point. The fixture is then adjusted until the pointer exactly touches the first point. This process is repeated with the second and third point.

Each of the reference points are checked and the fixture adjusted until the fixture position exactly matches points generated in the setup program. At this point, the fixture is located in the correct position and orientation with respect to the program.

For high accuracy applications, I have seen the reference points consist of precision bushings imbedded in the fixture. Instead of using a simple pointer, a sweep micrometer was mounted in the spindle head and the fixture adjusted until the micrometer indicated that the exact center of the busing was located. This same procedure was used to verify the fixture height.

If you require this level of accuracy, it is a good idea to construct a fixture base with fine adjust screws so that the position and height of the fixture can be fine adjusted and locked in position.

There are many variations of the many systems I have discussed here. There is no right way or wrong way to hold the part. If the part is held in the proper position and orientation, the fixture is right. If the part is not held properly or if the cut rattled or chatters, the fixture is wrong. Just remember, if you are having problems machining a part, the first place you should look is the holding fixture.

Chapter 4

Tooling

Tooling

There is an entire technology surrounding cutting bits. It is quite common for a new CNC router user to totally underestimate tooling requirements. In most cases, you cannot simply go the corner hardware store and buy a router bit that will perform properly in a production environment.

I have actually seen companies do this. They bought ten dollar bits because they were cheap. The tools slowed the cutting process in half and tool life was a small fraction of what a proper bit could offer. When the proper tool was used, which cost over fifty dollars each, production rate not only doubled but overall tooling cost was actually lower. The new tools lasted five to ten times longer than the cheap tool.

Now I am going to tell you an extremely important fact. Cutting tools have a limited life. Even though this seems obvious, I am forced to emphasize it because, believe it or not, some companies don't know this without being told.

I remember one instance where we sold a wood carving machine to a rather large company. We supplied a starter set of tooling and hoped to supply their ongoing tooling needs. We were disappointed when that didn't happen.

Six months later they called complaining that several of their spindles were running hot. Our service man determined that the problem was that the tools were extremely dull. In fact, they had run the original tools for six months, two shifts a day without ever changing them. We pointed out that you need to change tools when they become dull to which they replied, Oh!

Now that we are through the initial warning, the second thing to realize is that it requires a different tool for every material being cut. The material being processed will determine tool geometry, configuration and shape. There is no universal tool that will work properly on every material.

Improper tooling can cause several problems. Melt back, poor edge quality and surface scarring can result. Slow feed speeds or poor tooling life are also indications of improper tooling. Even excessive cutting noise can be caused by improper tool selection.

There is no substitute for a good, knowledgeable tooling supplier. If you can locate one that has had experience in cutting and machining the materials you plan to process, the advice can be quite valuable.

Be careful in this area, however. I have seen tooling salesmen claim that they have intimate knowledge and then proceed to supply completely wrong tools. I have seen salesmen blame the machine for poor quality edge finish when incorrect tooling was the real problem. In general, if the cutting process is not working properly, one of the first things to try is different tooling. If this doesn't work try a different tooling supplier.

I am not going to try and offer a complete analysis of the various tooling options available. Instead, I am going to try and cover the basics. I will also offer some general recommendations that we have developed after trimming and machining thousands of different parts in our demo labs.

Let us start with plastic materials.

When we are presented with a new part to trim in our customer demonstration department, we will start with an O-flute design tool and use a conventional cut.

An O-flute tool has a smooth, round relief to the cutting flute. This relief provides a path for the chips to peal away and exit the tool.

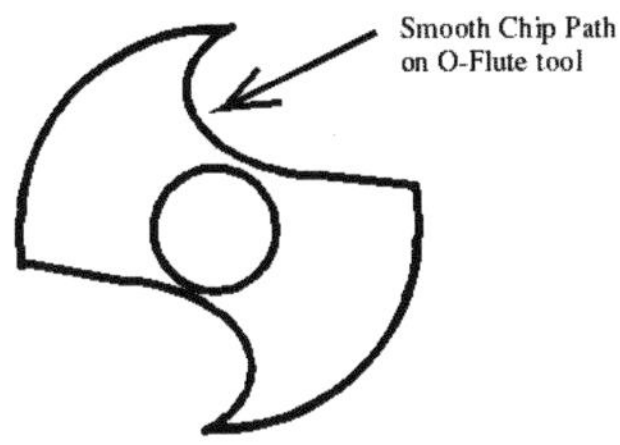

As this book is being written, this type of tool is being supplied by one vendor but several others have indicated that they will also offer similar tooling.

With a tool spinning clockwise as viewed from above, there are two different ways of cutting, climb cut and conventional cut. These are determined by the feed direction with respect to the material.

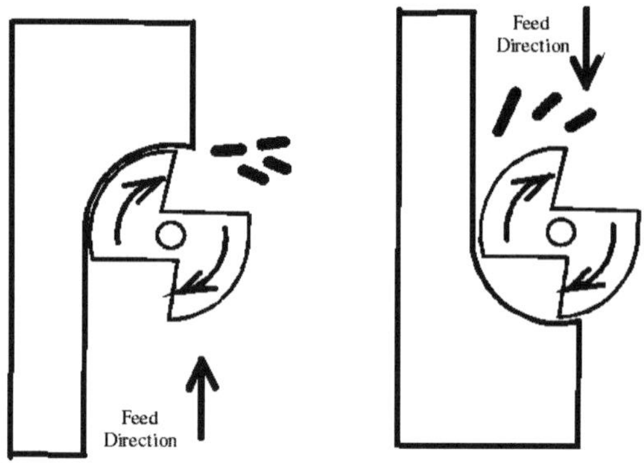

Conventional Cut Climb Cut

In conventional cutting, the tool slices into the material at the cut line and pulls the chips away from the part. This method tends to work well in cutting homogeneous materials such as plastic.

In climb cutting the tool slices into the part and removes the chips along the cut line. This cutting direction is commonly used to cut materials such as wood that can splinter using the conventional cut. In conventional cutting, the tool pulling away from the material exerts an outward force. If the material can split or splinter, this force may very well tear out a section of material. As you might expect this is called "tear out".

Tear out is not normally a problem in plastic machining and the conventional cut direction tends to offer a better surface quality.

Many times a cut is made through the material with material on both sides of the router tool. In this case, the type of cut (climb or conventional) is determined by which side of the cutter is the finished part edge and which side is the trim scrap.

The goal in plastic trimming is to obtain a smooth, clean, accurate trim edge at the fastest speed possible. To reach this goal, different tools and different feed speeds will be required for different materials and different thickness of the same material. Part shape and configuration can also have an influence on cutting tool selection.

The O-flute tool is available in several configurations. These include single, two and three flute designs in straight flute, up-shear, up-spiral, down-shear and down-spiral. Each of these is a specific geometric design that works in certain applications.

Unfortunately, each router bit manufacturer uses slightly different names for the various parameters that describe a tool. They each call their tools slightly different names. This means that it can be both confusing and difficult to compare one tool to another.

Cutting tool technology is advancing rather rapidly. This is not an old fixed technology as many people believe. Instead it is a rather fast moving science and I expect that over the next few years we will see continuing advances. This means it is wise to not only keep in close contact with your current tooling vendor, but also keep a dialog going with the other major tooling suppliers.

The most common problems encountered in machining plastics are melt-back, poor edge quality and surface scarring.

Melt-back occurs when the plastic trim chips remain connected to the top and bottom of the edge after the trim is complete. It looks like a rough plastic burr along one or both surfaces.

Melt-back is caused by the plastic trim chips becoming hot enough during the trim process to cause them to melt and reattach to the edge being trimmed. There are generally two causes for melt back. First, if the feed speed is too slow, friction from the cutting process is concentrated on both the edge and the very small chips. This heat can easily become high enough to remelt the chips and cause melt-back.

The second most common cause of melt-back occurs when the chips do not evacuate the trim area fast enough. If chips get caught up in the rotating cutting process, they can spin around with the bit. Friction from this spinning process heats the chips so that they then reattach.

The easiest way to identify a cutting problem is to examine the chips from the cutting process. The chips should look like the number "6" or "9". They should be thick on the inside and then taper off to the tip. If the chip is too thin or almost dust like, the feed speed is too slow. To correct this situation the feed speed must be increased until the correct chip shape is achieved. If the feed speed cannot be increased for some other reason, then the rotation speed of the router spindle should be reduced until the proper chip shape is achieved.

If the chip is thick and curved less than a "6" or "9", then the feed speed may be too high. This high feed speed is not necessarily a problem although it will normally lead to exaggerated tooling marks on the trimmed edge. If these tooling marks are not a problem, the higher feed speed has no other shortcomings. If a better quality edge is required, feed speed must be slowed or spindle speed increased until the correct chip shape is achieved.

At times the chip shape is correct but the surface finish is still not acceptable. In these instances, the problem is almost always the fixture and hold-down system and not the tool. If the part being trimmed is not held in place securely, it can move each time a cutter edge strikes it resulting in a rough finish. Sometimes this problem can be identified by a lot of noise during the cutting process. A high level of cutting noise or a rattling sound during the cut generally means that either the part is moving or vibrating or the cutting tool is incorrect for the material being cut.

Chip removal from the cut area is another concern. Sometimes a straight flute or a down-spiral tool can trap chips causing heat, melt back or poor quality edge finish.

An up-spiral tool will generally solve this problem by pulling the chips away from the part. The up-spiral shape, however, also pulls the part away from the fixture during the cutting process. For this to work properly, the part must be of sufficient stiffness and must be held rigidly enough by the fixture to prevent movement and the resultant poor edge finish. Sometimes, on thinner gauge parts, a down-spiral bit is used to push the part toward the fixture and assist hold down. In this case, however, chip removal must be watched carefully.

A straight flute cutter does not pull the part away from the fixture nor does it push the part toward the fixture. It therefore tends to trap chips less than a down-spiral but it does not provide any chip lifting should trapped chips become a problem.

To make a cut, first adjust feed speed to obtain a proper shaped chip. Listen to the sound of cutting. Working properly, you should have a smooth, controlled sound without chattering, proper shaped chips and a smooth finished edge. If a burr develops along the top and/or bottom edge of the cut, you need to increase feed speed or reduce spindle speed. If excessive tool marks occur on the part, you must slow the feed speed or increase spindle speed.

If you continue to have problems try variations of feeds and speeds. Some materials don't cut well at every speed. Also, if you are having problems with melt-back or surface quality and the parts are still warm from the molding process, it may be necessary to allow the parts to cool before machining. Some materials react quite differently when machined warm versus cold. I know of instances where different machining parameters were required to machine the same part made of the same material in the summer versus winter. The difference in the temperature of the factory affected the cutting parameters.

Tooling is an important part of the trimming process. Tooling technology is also changing rapidly. Remain open to suggestions from your machine manufacturer and tooling supplier but don't simply rely on their advice. My advice to you is to experiment. Try different bit designs, different feeds and speeds and determine what actually works and doesn't work in your facility.

Chapter 5
Programming

Programming

Programming is another area of CNC routers that is generally misunderstood. This is true even for people that regularly program CNC routers.

In this chapter I will try and discuss all of the different methods of creating the necessary programs. There are a variety of methods of developing programs each with advantages and disadvantages. To make things even more complex, many of these methods can be combined into an almost endless series of combinations.

I am not going to try and teach you how to program in this chapter. Learning to program a five-axis CNC router requires a book of its own. Instead, I am going to show you enough about programming for you to be able to understand the different methods used to create CNC programs in today's world.

I intend to explore this area in enough depth that you can understand subtle differences between specific code as executed by different controls when these differences have a material affect on the programming and use of the machine.

First, what is a CNC program? Today, a CNC program is a series of blocks of text. Each block of text is written in a specific format or structure that will be interpreted by the control. Each block of text will instruct the control to take some action.

Normally, each block of text is written on a separate line. On some controls line numbers will be shown, on others they will not.

The text is generally written in a format that is defined by an Electronic Industries Association standard. This standard is called EIA 274-D and it was initially intended to provide a consistent code format for NC controls built by different manufacturers.

When the standard was first developed, there was a great deal of concern over the standard stifling technical innovation. To keep this from happening, certain codes were reserved for future inclusion in the standard. In addition, certain codes were allowed freedom for each control manufacturer to define in any way they wished.

Control manufacturers have made extensive use of this freedom under the standard. In fact, they have made such extensive use of this freedom that few, if any, controls manufactured by different companies are code compatible. This means that code written for one machine will generally not run on another machine built by a different company. In fact, in some cases code written for one model machine will not run on another model machine even if controls for both machines are built by the same company.

NC code is sometimes called “M” and “G” code because each block of text normally starts with the letter “M” or the letter “G”. The letter “G” normally prefixes a machine motion line of code and the letter “M” generally prefixes an input, output or tooling control line of code.

There are two different methods of specifying axis motion, absolute and incremental. When the control is in the absolute mode, an axis position, X10.5 for example, specifies a position with respect to the home or 0,0 position of the axis. In this example, the position specified is 10.5 inches in the positive direction from the home position on the X Axis.

When the control is in the incremental mode, the code specifies a distance and direction to move the axis FROM THE CURRENT POSITION. Thus, the same code, X10.5 will produce a totally different motion depending on whether the control is in the absolute or incremental mode. In the absolute mode it will move to position 10.5 on the X Axis and in the incremental mode it will move 10.5 inches in the positive direction on the X Axis from its current position.

The absolute and incremental mode can be set using a "G" code. These codes are "modal" which means that whenever they are set, they remain active until changed.

There are many modal commands that remain in effect until reset. The speed of an axis, for example, is normally a modal command. Once the feed speed is set it will remain at that level unless reset. A good control will display the modal commands that are active at any time.

It is possible to switch between absolute and incremental modes within a program, but this can cause problems. It is very important to choose one mode that is normal, incremental for example. Then, anytime the absolute mode is used you must make certain that the code returns the control to the incremental mode before moving on or branching to another location.

A mistake in this area is a common reason for crashing a machine into a fixture. For example, let us assume we are at the 55 inch absolute position on the Y Axis and want to program a line of code to achieve a 2 inch motion to the left. The code for this is G1Y2. "G1" indicates it is a linear motion and "Y2" indicates the motion is 2 inches in the positive direction on the Y Axis.

If the control is in the incremental mode it will, indeed, move 2 inches to the left. If, however, the machine was inadvertently left in the absolute mode, the axis would move 53 INCHES TO THE RIGHT! You can see how easy it is to crash the machine if you don't handle incremental and absolute modes properly.

G00 is the command for a rapid traverse linear motion. For example the line:

G0X10.5Y5.65

is a line of code that tells the control that we want a rapid motion with the X Axis moving 10.5 inches in the positive direction and the Y Axis moving 5.65 inches in the positive direction (we will assume the incremental mode from here on unless noted).

The results of this command will be different on different machines. Under EIA 274-D, the G00 command does not require linear interpolation. Linear interpolation is a feature where the control coordinates the motion of two or more axes so that they start and stop at the same time.

In this example, with linear interpolation, the machine will tract a diagonal from the current position to a position 5.65 inches to the right and 10.5 inches forward of the last position. The path of the motion will be a straight line from the last position to the next.

Without linear interpolation, each axis is instructed to move as fast as possible from its own individual position to the next position without regard to what the other axes are doing. In this case, both the X and Y Axes would begin moving at the same speed until each had moved 5.65 inches. Then the Y Axis, which has already achieved its new position, would stop moving and the X Axis would continue until it moved a total of 10.5 inches.

The path traced would be a 45 degree diagonal for 5.65 inches along both the X and Y Axes and then an additional 4.85 inches along the X Axis. Why would you want to do this?

Actually you don't. Linear interpolation requires sometimes complex calculations to assure coordination between the axes that are moving. Early controls were too slow to perform these calculations at high feed speeds. The machines were required to move rather slowly so that the control could keep up the necessary calculations.

At times, however, it was desirable to move to another location quickly. Normally this was to re-position the head for another cut. The G00 command was used to free the control from the necessity of coordinating the axes so that higher feed speeds could be achieved.

With today's high-speed processors, coordinating the axes is no longer a limiting factor. On the other hand, not coordinating the axes on a five-axis machine can cause serious problems.

With coordinated or interpolated motion, you know the exact path that the machine will follow. Without interpolation, the path can be unpredictable. This unpredictable path is not a problem with a simple two-axis machine. The head is simply moved above the work and the path taken to the next position is not important. In this case there is nothing to crash into.

With a five-axis CNC router, many times the fixture is so high that you cannot get over the top of it. In this case, the path is very important. It would be quite easy for a non-interpolated positioning move to crash into the fixture. It makes no sense to have a G00 command that is not interpolated, even though interpolation is not strictly required by the standard.

Why have a G00 command at all?

Actually, the G00 command can be quite useful. This command is used to reposition the machine between cuts. Since we should not cut using this command, we can move as fast as the machine is capable of moving. In fact, the acceleration and deceleration rates can be increased beyond the point where smooth cutting motion occurs. This means faster motion between cuts and shorter overall cycle times.

Also, another feature of most controls is called "feed speed override". This normally consists of a knob that will either slow down or speed up the programmed feed speed on the machine while it is executing a program. The operator can use this knob to fine tune the feed speed to the actual material and cutting tools.

The G00 command is not affected by the position of the feed speed override control. Slowing the execution of the actual cuts will not slow the G00 repositioning motions.

The G00 command can be useful and will result in faster, more efficient programs. Make certain, however, that the control you are using provides full interpolation for this command even though it isn't an official requirement for the command.

I am not going to detail each of the "M" and "G" codes in this chapter but instead will focus on the various methods of creating programs that use these codes.

The most obvious and simple way of creating code is to sit down at a word processor and type each line of code required. Simple and easy, however, are not the same thing. Although this method creates code in a straightforward manner, it is all but impossible to envision the position and orientation of a five-axis machine while working with a word processor.

This simple method can, in fact, be used to program straightforward two-axis motions. Even then, since the diameter of the tool must be taken into account, it is not particularly easy. Once the part requires three or more axes, it is all but impossible to create program code directly.

Perhaps the easiest way of creating program code is programming right at the machine, where the machine moves as the program steps are developed. In this way, you can see the resulting positions and paths and make any necessary corrections immediately.

This approach has other advantages. As part of the programming process it allows for the exact location of the fixture on the tabletop. It also allows for any positional inaccuracy of the machine itself. We will discuss accuracy in detail later in the book, however, even a relatively inaccurate machine can produce consistent parts as long as it repeats accurately. When the machine is used to create the program, any machine inaccuracies are allowed for in the program. The only real disadvantage is that a program that allows for positioning inaccuracies of a machine will only function properly on that one machine. Programs of this type cannot be moved from one machine to another.

At this time I am going to introduce you to the Hand Held Programmer. This programming terminal was introduced by Thermwood in the late 1970s and remains a very powerful and popular programming tool.

The Hand Held Programmer looks like a large calculator. It has a keypad, a two line display, a rotary positioning knob and an emergency stop button. It is plugged into the front of the control and is connected to the control by an umbilical cord. The Hand Held Programmer can be taken right to the machine where programs can be developed while you watch the results from close range.

The basic concept behind the Hand Held Programmer is quite simple. Just imagine a simple milling machine that uses hand cranks to move the table. Now imagine that you move the table to a position and then press a button to store that position. Then you move the table to a second position and again press the button to save that position. When the program runs, it moves from one saved position to another.

This essentially is how the Hand Held Programmer works, except that instead of using the hand cranks to move the table we use the keys on the Programmer. To try and understand this a little better, let us imagine we are creating a program.

Let us assume we need to trim around the base of a simple thermoformed part to remove the excess flash material. The part is more or less rectangular, but the trim line is not totally horizontal. This means that we will need to approach the part with the head tilted up somewhat. We have the part mounted to a fixture located somewhere in the center of the table.

We will use a ¼ inch diameter router bit about an inch and a half long to trim the part. The first step is to move the head to the initial plunge point on the part. We will use the Hand Held Programmer to move the machine.

First select the axis you want to move. Type ***Axis*** then ***1*** to select the X Axis. To keep the number of keys to a minimum, numbers are used to designate the axes on the Hand Held Programmer. 1 is the X Axis, 2 the Y Axis, 3 the Z axis. 4 and 5 are the two rotary head axes.

Now that we have selected the axis, type in the distance you want to move, 20 inches for example. Press the + key, indicating you want to move in the positive direction which is away from home on this axis. At that point the X Axis of the machine will move 20 inches. Now select ***Axis 2*** (Y

Axis), type in a distance and press +. The Y Axis will move. Do the same for the Z Axis. We are now getting the head closer to the part.

We will now orient the head so that the router bit is pointed in the proper direction. This is done in the same manner. Select Axis 4 and Axis 5, type in the amount of rotation and the head will move. On axis 4 and 5 however, the units it will move are degrees and not inches. 90 degrees will move Axis 5 from horizontal to vertical.

At any time we can return to any axis we have already moved and change its position by adding to or subtracting from its position. The display will indicate the active axis and the distance we have moved so far.

We now have the head oriented properly and are only a few inches from the part. What we really want to do is to position the tool about an inch or so away from the part and then plunge into the part along the axis of the tool. When the program is started, the head will simply move away from home, position itself away from the part and then plunge straight into the part and begin trimming.

From an inch away from the part it will be difficult to judge the exact plunge point so, instead we will move the tip of the tool right to the plunge point and then back it off an inch. This assures us that when we plunge it will be in the proper location.

Since we are getting close to the part we will move in smaller increments. Type ***Axis 1 .25*** and press +. The X Axis will move .25 or ¼ of an inch. Press + again and it will move another ¼ of an inch. Each time you press the + or –

key the axis will move ¼ of an inch since that was the last dimension you typed in.

Select ***Axis 2*** and type ***.5***. Press +. The Y Axis will move ½ inch. Again, each time you press the + or – key the axis will move ½ inch. Now return to the X Axis and press +. It will move ¼ inch, not ½ inch. Each axis has its own default distance, which is the last distance you typed for that axis. Changing distance on another axis will not affect it.

Although this may sound a little complex, within just a few minutes it becomes quite natural and comfortable.

As you approach the part, there is another way to move the axis. There is a pulse wheel located at the bottom of the Hand Held Programmer. Turning this wheel clockwise will move the axis in the plus direction and turning the wheel counterclockwise will move the axis in the negative direction. As you rotate the pulse wheel you can feel individual detents at about one-degree increments. The distance that the axis moves for each of these detents can be adjusted by a series of keystrokes on the Hand Held Programmer.

We are now positioned with the tip of the router tool just touching the plunge point on the part. Now all we have to do is back off an inch. This however, is not quite as simple as it seems.

Since the head is positioned at an angle, we will need to move all three linear axes, X, Y and Z a precise distance to move the head along a line through the center of the router bit. How much do you move each axis? Actually, there are two ways to approach this problem. The first one is to find somebody that is good at trigonometry and have them calculate the exact distances to move each axis. The current rotational positions of each of the rotary axes can be read off the Hand Held Programmer display.

A much easier way, however, is to simply call up ***Axis 0*** on the Hand Held Programmer. Axis 0 is really not an axis, but instead it is an aid in positioning the head. When Axis 0 is called and a distance typed, the machine will move all three linear axes so that the head moves along the centerline of the tool whatever distance was typed. In this case, + is toward the part and - is away from the part.

All we need to do is type ***Axis 0, 1 –*** and the head will move one inch away from the part along the centerline of the tool. This is the position we have been searching for all along. All we need to do now is press the ***ENTER*** key and the control will write a line of NC code to move from the home position to a position one inch from the part with the head properly oriented.

There are a few things we should note at this point. First, the steps necessary to locate the head are more complex to describe than to actually do. An experienced Hand Held Programmer can actually locate the head properly in less time than it took you to read about it. Also, the first motion that you programmed was a full five axis simultaneous move. You would not be able to create this motion on any control that restricted the number of simultaneous axes that can be programmed. Instead, you would need to break the motion down into two or more individual moves that eventually ended up in the same position and orientation. If you plan to program at the machine, make sure you have both a Hand Held Programmer and full five-axis simultaneous motion or the task becomes much more burdensome.

Let us return to programming our part. We are now an inch from the initial plunge point ready to move into the part and begin trimming. Let us consider exactly how we are going to program the rest of the part.

We either have a fully trimmed part or we have an untrimmed part where the trim line has been clearly marked. The approach for each of these is different.

If we have a fully trimmed part and if a groove has been machined in the fixture to provide clearance for the router bit, we can actually program to the final path. If the part is not trimmed or the fixture has not yet been relieved, we will need to program in front of the path and then move the cut path into the part when we are ready to actually machine the piece.

The easiest way to accomplish this is to slide the router bit all the way in, or use a shorter router bit for programming and then replace it with a longer bit when it is time to cut. On some controls, five-axis tool length compensation is available. This allows you to shift the program, in all five axes, automatically to allow for different length tools. If your control has this feature, it is not necessary to play around with tool lengths. Simply create the program by moving the tool around the cut path, but slightly off the part, then use tool length compensation to shift the tool into the part for cutting.

As you define more and more movements, you can see how these will look as they are executed. Press – then ***JOG*** and the machine will move back to the last point you programmed along the path you programmed. Press – and ***JOG*** again and it will move back yet another step. Press ***JOG*** alone and it will move forward through the program, one step at a time. Using this feature makes it quite easy to edit and refine the program while it is being developed.

There are many other features and techniques that have developed around the Hand Held Programmer. Key sequences are available for creating circles and arcs. Any three points can be turned into an arc, including free arcs in space.

At times it is necessary to execute a smooth flowing curve along a part. It would be quite difficult to program a large number of points very close together using the Hand Held Programmer and achieve a smooth, even motion. To address this requirement a capability from CAD/CAM systems has been added to the Hand Held Programmer toolbox. This capability is called a spline.

A spline is a mathematical function that takes a number of points and creates a smooth path through them. There are a number of different types of mathematical splines, but this can become much too involved for a discussion here. For our purposes, consider a spline a function that creates a smooth path through a series of points.

Splines initiated in boat building. The shape of the keel and hull of a boat were designed and small nails were driven into a wall to define the shape. A long, thin strip of wood about an eighth of an inch thick was then bent over the nails to define the smooth shape of the hull. This strip of wood was called a spline.

As the boat design process became more refined, mathematical functions were developed to replace the wood splines in defining a smooth curve through a series of points. These mathematical functions took on the name of the wood strip they replaced, the spline. As CAD/CAM systems developed, these mathematical functions became important tools for defining smooth curves and surfaces.

Thermwood has added spline capability to its control system in a couple of different forms. When you want a smooth curve through a series of points on a Thermwood control, you simply place a spline marker at the beginning of the points and a spline end marker at the end of the path. When executed, the control will calculate a spline, real-time, through the points.

Spline calculations are extremely involved. Even with today's very fast processors, they take time. For this reason, top speed when executing a spline that is being calculated real-time is limited. If you wish to cut at a higher speed, there is an alternative. A program with spline markers can be run through a utility called "CAD Path". This will generate a program where the spline is expanded in the same manner as a CAD/CAM system would expand a spline path. This expanded path can then be executed at any feed speed up to the limit of the machine. The "CAD Path" utility is part of Thermwood's 91000 SuperControl. I am not aware of any other control that has spline capability but there is no technical reason that it could not be added to these systems and I expect that they may be available for other systems in the future.

Spline capability and CAD Path bring CAD/CAM type programs to companies that create programs using the Hand Held Programmer. It does this without the need for a full CAD/CAM package, computer hardware and trained CAD/CAM programmers.

I have spent some time describing Hand Held Programming features. I have done this so that you know what to expect from this tool. Many, if not the majority of companies programming three-dimensional trim parts today program exclusively with the Hand Held Programmer. It is a very powerful tool and can be learned by almost anyone.

Because of their popularity, some CNC router manufacturing companies have introduced their own “Hand Held Programmer” that is not much more than a utility box and a feed knob. If you plan to program at the machine, make certain that the programming terminal, Hand Held Programmer or whatever they call it has all the features you will need to do a quality, efficient job of creating your programs. If this is to be your primary programming tool, make certain it is a good tool.

Now that you have had a glimpse of how the Hand Held Programmer works, you may have noticed something. It seems that the majority of programming time is spent trying to get the machine and head in the proper position. Most of the keystrokes are used to try and locate the desired point. Wouldn’t it be nice if we could simply guide the machine into position by hand and then enter the point? Now you can.

In 1997, Thermwood introduced a programming probe which does essentially that. The probe mounts to the router spindle where the router bit would normally be located. The probe is about six inches long and four inches in diameter and will make the router head about six inches longer than normal. This offset will be removed using tool length compensation so that the actual router bit will travel the path defined by the probe when the resulting program is run. The extra length, however, will reduce the overall working envelope of the machine so parts that challenge the capacity of the machine may not be able to be programmed using a probe.

The probe is used by inserting a probe tip into the probe body. The probe tip is the same size and shape as the router bit we intend to use. This probe tip is then used to guide the machine around a part.

The probe tip moves slightly from side to side and in and out of the probe body. These movements are translated by the probe into electronic signals that are fed to the machine control. These signals are used to move the three linear axes of the machine. What happens next is a bit complex, but first let us look at how we move the two rotary head axes.

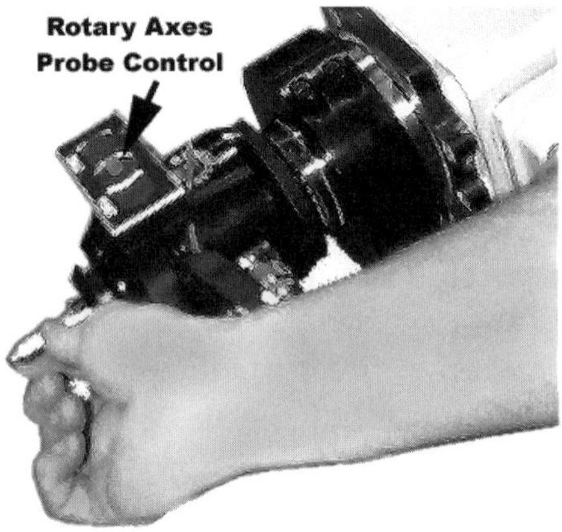

Located on the probe body is a small red “joystick” button, like those found on some laptop computers, that is used for rotary axis probe control. This button controls the two rotary head axes, Axis 4 which rotates the entire head horizontally and Axis 5 which rotates the router spindle on the support arm. As you pull down on the button, the servo drives rotate Axis 5 and the spindle downward. As you pull up on the button, the head moves up. It feels like you are pushing the head using the button.

Axis 4 is a little more complex. When the head is positioned with the button up, as shown, pushing the button to the right causes the head to rotate to the right. Pushing the button to the left causes the head to rotate to the left. Again it feels like your are rotating the head using the button.

When Axis 5 rotates the head so that the button is down, the sense of the button changes. Again, when you push the button to the right the head rotates to the right and when you push the button to the left the head rotates to the left. What you may not realize is that when the head is positioned with the button up, the direction that you push to cause Axis 4 to rotate clockwise is the same direction that you push the button to cause Axis 4 to rotate counterclockwise when the button is down. Thus, the same signal from the button causes two completely different actions on Axis 4 depending on the position of Axis 5.

As I said, this is a little complex. In order to get a natural feel when controlling the head using the button, the control must determine the sense from the overall position of the head and the direction you are pushing the button rather than simply moving the machine in response to a signal from the button.

If the button is pushed at an angle, both Axis 4 and 5 will rotate at the same time.

Now we can look at the three linear axes. Controlling the three linear axes with the probe is even more sophisticated. Let us assume that the head and probe are positioned along the X Axis. Pulling on the probe tip will then move the X Axis. If we now rotate the head so that the head and probe are orientated along the Y Axis, we want the Y Axis to move when we pull on the probe. In one orientation we want the signal from the probe to move one axis and in another orientation we want that exact same signal to move a different axis.

When the head is positioned between the two axes, we want to move both axes. The rate that each axis must move with respect to the other, in order to get a natural "feel", must be proportional to the angle of head position at the time the signal is being received. When we rotate Axis 5 we must now move all three linear axes when the probe tip is moved in order to get the proper action.

As I said, the technology involved is quite complex, but the final result is that regardless of the orientation of the head, the machine will move in any direction that you push or pull the probe tip. In fact, you can simultaneously move the three linear axes while rotating the two rotary axes and the system will constantly change its reaction to the probe inputs to provide a simple, natural feel to the motion.

There are three modes in which the probe can be used. The first of these is the "point mode". In the point mode, the probe is used to move the machine axes to a point much like the Hand Held Programmer was used to move the machine to a desired point. When the point is reached, a record button on the probe is pressed and the control creates the line of NC code necessary to move from the last recorded point to the current position. Move to the next position and press the record button again. This process is continued until all the points necessary have been recorded.

As you can see, this process is much faster than recording the exact same points using only the Hand Held Programmer. By the way, the Hand Held Programmer can be plugged in and used at the same time that the probe is being used. For example, the probe could be used to locate a plunge point. Then the Hand Held Programmer, using the Axis 0 capability could be used to back the head away from the part, record the point and then return to the plunge point. The probe could then be used to continue programming. In another example, the probe could be used to locate the edge of a circle. The actual circle can then be programmed using the Hand Held Programmer. A circle programmed with the Hand Held Programmer will be much more accurate than one developed with the probe alone.

As you can see, in the point mode, the probe is an extremely useful tool that makes program development much faster and easier.

How does the machine know the position of the probe tip? This is not as simple a question as you might think. First, the control can look at the position of all axes and determine the exact location of the machine head. This, however, is not the position of the probe tip. As you recall, the probe tip moves slightly to provide the input signal to generate machine motion. If we simply use the position of the head as defined by the axes positions, we are assuming that the probe tip is exactly centered. In actual practice, it may not be perfectly centered when we record a point.

Since we are using the probe tip to define a position, we must determine the exact position of the probe tip and not the position of the machine axes.

The probe system that Thermwood uses has three very accurate linear scales that measure exactly how far off center the probe tip is in three perpendicular axes inside the probe body. This offset information is sent to the machine control. The control then takes the position of all machine axes, the offsets from the probe and the rotational position of the two rotary axes and calculates the exact position of the probe tip. As you can imagine this is a rather involved calculation.

I have included some information on exactly how the probe works because most people that use the probe find it so simple that they totally underestimate the underlying technology.

The second basic mode in which the probe operates is called the "drag mode". The drag mode is used to define a curved path in space using only the probe. There are actually a number of different ways of programming a free curved path in space. The method that is best depends somewhat on the application and requirements.

The most straightforward method of creating this path is by using the drag mode and the programming probe. To use this method the probe is moved to the beginning point of the path we wish to program. This is normally done using the point mode. At this point we activate the drag mode by pressing the drag button on the probe.

The goal in programming a smooth curved path in space to create a program that consists of a series of very short line segments that connect to one another and define the path. If these segments are short enough without abrupt changes in direction, the resulting program will appear to be a smooth, continuous curved line in space.

To achieve this smooth path, typically the line segments are quite short, ten to twenty thousandths of an inch for example. Because the segments are short and there may be thousands of them, it is not possible to enter these segments one at a time.

In the path mode the probe creates this type of program. The way the probe works is a bit less direct, however. When you activate the path mode, the probe begins recording the position of all axes very quickly. Although this record interval can be adjusted, it is normally set at about six milliseconds. Every six milliseconds the control records the position of all axes whether the axes have moved or not. During this time, the operator is tracing the path using the probe. This tracing process is generally done slowly and carefully.

When the path mode is turned off, the actual program code is developed by the control. The control now has a stored list of the position of each axis throughout the tracing process. To convert this list into a NC program, the control creates a block of code for the start point of the path. It then begins calculating the distance that the tool tip moved for each set of axes positions that has been recorded. When the combined movement is equal to the segment length, the control again writes a line of code to move from the last point to the current point. The length of this segment length can be specified. It then discards all the points used to reach this position and begins the process again, defining another segment. It continues the process, using the recorded axes positions to define set length segments until the entire path has been programmed.

The result of this process is a program that consists of a series of generally equal size segments connected together along the desired path. The program is essentially the same as a program generated by a good quality CAD/CAM system.

In practice, programs where the head orientation is set and the three linear axes are used to create the program path work extremely well. Defining a complex path in space using the probe and only linear axes is easy. If a rotary axis motion is added to the linear axes during the segment, it becomes a bit more difficult to get the smooth, professional program you are looking for. The probe and control are more than capable of handling input from all five axes simultaneously. The difficulty that occurs is with smoothly coordinating the linear and rotary axes by hand. Some people are able to do this rather easily while others never do seem to get an acceptable path. Luckily, there is a simple way of achieving a virtually perfect path that combines both linear and rotary axes.

We will now go back and record a series of points along the path using the point mode. In this way, all the axes can be positioned slowly and carefully using the probe, and if necessary the Hand Held Programmer. Depending on how quickly the orientation changes, these points can be anywhere from a quarter of an inch to an inch or more apart. After the points have been entered, a spline marker is entered in front of and at the end of the line segment.

After the program has been completed, it will be processed through a utility in the control called CADpath. CADpath will create another program where the points between the spline markers are converted into a CAD/CAM type path with short, connecting line segments. The final result is essentially the exact same output you would get from a full blown CAD/CAM system, yet the process can be completed by a machine programmer that is not trained in the operation of a complex CAD/CAM computer system.

The reason that CADpath does not modify the current program is in case you need to modify or edit the path. Should you find that a point or two needs to be adjusted, it is quite easy to take the original program with a limited number of original points and adjust it. Then, simply run it through CADpath again and you have an edited version, quickly and easily. If we had modified the original program, there may be thousands of points and editing would be much more difficult.

With CADpath it is easy for almost anyone to create smooth professional programs that exactly match the part. The high tech world of CAD/CAM programming has been made simple enough that almost anyone can do it.

The third mode that the probe can execute is the scan mode. The scan mode is used to create a program to reproduce a surface by making a cut, stepping over a slight amount and returning. The machine runs back and forth over the part, stepping over a little bit each time until the entire surface has been cut.

The probe and the scan mode can be used with a physical model to create the necessary program. To operate in the scan mode, the perimeter of the part to be scanned is defined and the step-over distance is specified. The scan process is then started.

The machine and the probe will automatically move back and forth over the model, accurately tracing it while creating a meta-file. The meta-file is much like the list of axes positions that we discussed earlier. It contains the axes positions recorded every few milliseconds. The logic and motions needed to trace the part will vary depending on how radical the surface variations are. If steep walls must be climbed or deep valleys followed, the movement speed of the tracing process is automatically slowed. Because of the rather involved calculations necessary to convert these paths into program code, it is much faster to simply copy the axes positions and process it than to try and both control the machine motion and process the program at the same time.

There are some limitations to the surfaces that can be scanned, but generally, most parts that are to be produced in a simple closed mold without undercuts can be scanned and cut using the programming probe.

The programming processes we have just covered are generally done at the machine. This has some major advantages since the actual part and actual machine are used to create the program, generally accuracy of the trim path is quite good, even for machines that are not particularly accurate or well aligned. To function properly, the machine must be repeatable, but if the part is programmed at the machine, the machine does not have to be accurate. There is a major difference between accuracy and repeatability and we will cover this area in much more detail later in this book.

In the past, few companies turned to CAD/CAM systems for program development because they were complex, cost millions of dollars and really didn't work very well. That is all changing today.

CAD/CAM systems are available at a variety of prices, can operate on reasonably inexpensive computers and are quite capable today. In the next few pages, I am going to try and offer an overview of systems available today including costs and technical requirements. Since this is a rapidly moving field, some of this information will become obsolete rather quickly. You might want to note the publication date of this book to determine just how far behind it might be at the time you are reading it.

The basics of CAD/CAM will, however, generally remain constant so much of what I will talk about should still be accurate regardless of the publication date.

Before we discuss specific CAD/CAM systems, I would like to explain the fundamentals of how a traditional CAD/CAM system works. From an understanding of the basic operation, we can then discuss the various modifications that certain system vendors have made to the basic operation.

CAD stands for **C**omputer **A**ided **D**esign. CAM stands for **C**omputer **A**ided **M**anufacturing. Although the two sets of letters are commonly used together, they represent two different functions. Let us start with CAD.

The CAD portion of the package is used to design the part. To help us understand this, we will explain how a very simple two dimensional part would be designed and programmed using a CAD/CAM system.

For our example we will make a square with rounded corners like this:

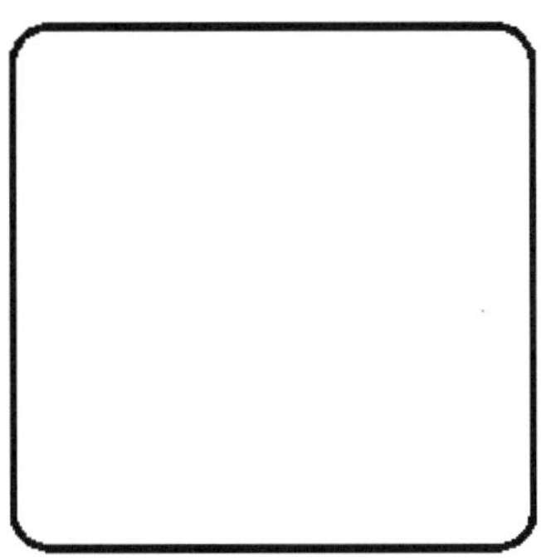

The first step is to design the part using the CAD system. CAD systems all have a simple system for creating a square and specifying its size. We now have a design that looks like this:

Now we must round the corners. CAD systems have a command, normally called the “fillet” command where you can specify a radius and then simply point to the junction of two lines. The square corner between the lines will be replaced with a curve, with the specified radius that smoothly blends the lines together.

Applying the fillet command to each corner gives us our part, a square with rounded corners.

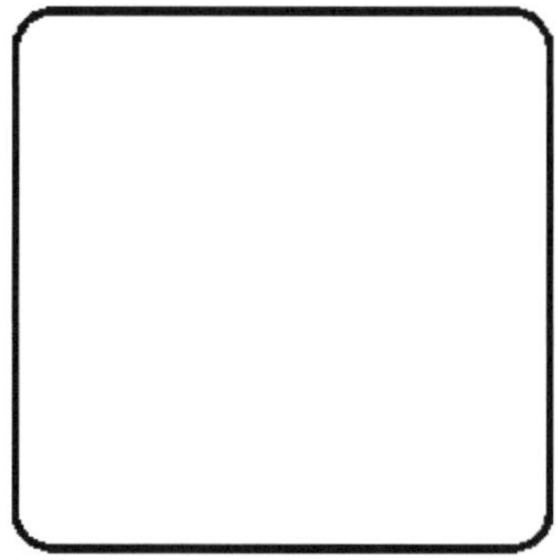

We have now designed the part. Depending on the CAD system, this information can be saved in a number of different formats. Some of these formats are used by multiple systems and others are used by only one package. If your CAD system can use one of the standard formats, the AutoCAD DXF format for example, then the design you create can be moved from one system to another. This is a useful feature even if you only plan to use one system. Should you decide to use outside programming help or want to upgrade your system in the future, you will be happy you selected a system that is generally compatible.

At this point, the work of the CAD system is done. If we want to develop a NC program to manufacture the part on a machine, we need a CAM system. Let me explain the things we need to do to the existing part design to create a machine program.

If you recall, we designed the part by creating a square and then filleting the corners, one at a time. The design data will be made up of four lines and four arcs. Since we created the square first, the four lines will appear first on a list of design features. The order of the lines on that list will depend on which direction we created the square. The four lines will be followed by four arcs. The order of the arcs on the list is determined by the order in which you specified the fillet commands.

If you simply use this list of design features to machine the part, the machine would first machine the four lines, one at a time and then it would cut the four rounded corners. The direction of cut would not necessarily be correct. This is not what we want.

To properly cut the part, we want to start at a point off the part and move into the part cutting in the proper direction. We then want to move smoothly around the part, first cutting a line, then the arc, then the next line until we traveled completely around the part one time.

To accomplish this, we will need to rearrange the order of the design features that we programmed. We may also need to swap the direction of some of the lines or arcs so that they are drawn in the direction we want to cut.

This is the first major job of the CAM system. It orders the geometry and cut direction and links the geometric entities together. We now have a smooth continuous line around the part in the correct direction, made up of the lines and arcs we developed in the design process. Can we now use this to program the machine?

Not quite yet. The lines we have represent the edge of the part. A machine program, however, defines the path taken by the center of the cutting tool. If we try and use the lines we developed as the cut path, the diameter of the cutting tool will make the part too small.

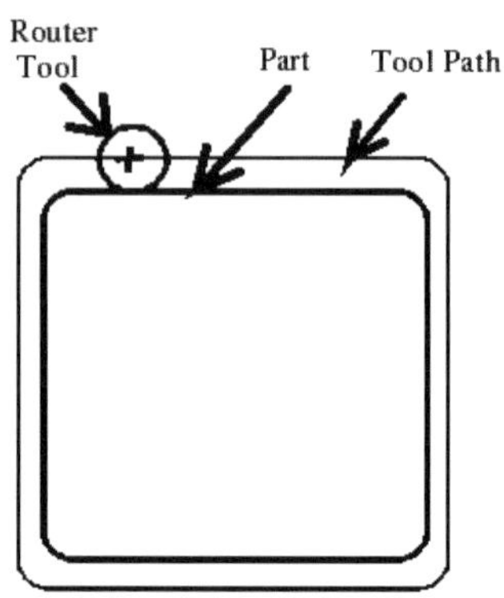

We must shift the cut path away from the part an amount equal to the radius of the cutter. In this way, the edge of the cutter will trace along the edge of the part. This is another major task for the CAM system.

Each CAM system works a little differently, however, they must all reorder the data, set the cut direction and offset the cutter. Once all this has been done, the result is generally known as center line data. It is a set of data defining the path that the center line of the tool is to take. It is still not a machine program.

The final part of this package is known as the post processor. The post processor is a program written for a specific machine that translates the center line data into NC code. We now have a program that will run on the machine.

Post processors can be quite simple for a two-axis machine or can be extremely involved for a full five-axis machine. On a five axis machine, the post processor must not only know the syntax of the NC code required by the control, but must also know the mechanical configuration of the machine itself. The center line data defines the path that the tool must take. The post processor must then calculate the movement of all five axes required to achieve the defined path with the tip of the tool. This makes a five-axis post processor a rather involved program.

Here, in a very simplified form, is the operation of a CAD/CAM system. The CAD design package designs the part and outputs geometric data. The CAM package creates a file that defines the path required for the center line of the tool to machine the part and a post processor converts the center line data to an NC program that can run on the machine.

Now that we have some idea of how a CAD/CAM system works, let us look at the various types of CAD/CAM systems available.

In general, CAD/CAM systems can be separated into two major categories, those that work with two-dimensional parts and designs and those that work with three-dimensional designs.

Obviously all parts are three-dimensional. From a CAD standpoint, however, a two-dimensional part has its design features defined by the X and Y axes. The Z Axis or part thickness is generally a fixed distance with Z Axis motion being simple plunge and retract motions.

We will begin our discussion by looking at systems for designing simple flat parts with a three-axis CAD/CAM system. These systems are called three axis CAD/CAM because they are intended to program three axis machines. In general they will develop path data for two of the three axes and will provide plunge and retract data for the third axis.

There are a number of relatively low cost three-axis systems available in the market today. They work essentially the way we described earlier using our square with rounded corners.

There are also a number of CAD-only programs, such as AutoCAD, that can design parts. These are not full CAD/CAM systems since they do not have the programs needed to perform the CAM functions. There is, however, a method of using these systems as a full CAD/CAM package by using a unique feature of Thermwood's 91000 SuperControl.

This feature, called a DXF translator, will work with any CAD system that can output its geometry in an AutoCAD compatible DXF file format. DXF is a file format for CAD generated geometry that was first developed by AutoCAD. This format became widely accepted and many CAD systems have the ability to output files in the DXF format.

To make the Thermwood DXF Translator work, the CAM data is added to the CAD drawing using text statements. To make the system work, we must specify the location of the plunge, the direction of cut and which side of the line the cutter is on. To accomplish this, a text statement in a fixed format is written across a line in the drawing. The endpoint nearest the text statement will be the plunge point. The direction of cut is the direction from the endpoint toward the text statement. The location of the text statement on the drawing has already provided two of the needed pieces of information.

The text statement itself contains a sequence number in case more than one path is to be cut. It also contains a "G" code indicating which side of the line the cutter is to be offset. At this point, all the necessary information is available.

When the file is run through the DXF Translator in the Thermwood control, the control will automatically reorder the geometric data and create the appropriate program to machine the part. It is also possible to add any other commands to the text statement. The control will execute these commands before executing the actual path motion.

This system allows relatively low cost CAD systems to operate as full CAD/CAM systems when a two-dimensional tool path is all that is necessary. When we need to program and run a full five-axis machine, things get a little more complex.

There are two basic machining functions for a five-axis CNC router. The first is to trim a molded part along a trim line. This can include cutting out opening in a part as well as removing flash from a molded part.

The second basic function of a five-axis CNC router is to machine surfaces such as those found on patterns or molds.

While the same CAD/CAM systems are used for both of these functions, the methods that are used and the features required of the system are somewhat different for each application. We will first address the part trimming application and then talk about programming and machining surfaces.

When we consider trimming a part, we will need to match a trim program to a real world part. This can be more of a problem than it appears on the surface.

To program a trim line using a CAD/CAM system requires that either the part geometry or at least the trim line geometry and head vectors be available. It is a common error to assume that if the original part was designed using a CAD system, the necessary geometry already exists.

In plastic molding today, almost never does the actual part size exactly match the size developed using the CAD system. Variations in pattern making, mold making, part shrinkage and machine absolute accuracy result in parts that are close, but not exactly what was intended. In most cases, these variations are significant enough that if the original CAD design is used to generate the trim line, the resultant trimmed part is not commercially acceptable.

This means that, in most cases, you will need to somehow modify or replace existing part geometry data with data describing the actual part size before an acceptable trim path can be developed using a CAD/CAM system. There are two methods I am aware of for doing this. The first is by using some type of coordinate measuring machine to measure the actual part. These machines, although quite expensive, can very accurately measure a trimmed part and provide most CAD systems with the necessary points needed to generate accurate geometry. This will correct for all accumulated errors that occurred up to the point where the part was complete and ready to trim. It will not account for absolute positioning errors in the actual trimming machine itself.

Another less costly but somewhat less accurate method of obtaining real world points is to use a three-dimensional computer input device. These input devices are normally jointed arm designs that input to the computer the three Cartesian coordinates of the tip of the arm.

A trimmed part is placed near the arm and the tip of the input device is used to pick up points along the trim path. It can also be used to pick up points on a surface so that the CAD system can recreate the surface.

Many times all points on a part cannot be reached by the arm without moving the part. In this case, reference points can be established on the part, then the part moved and the reference points again located. Software within the system will then blend the points together. In this way, a part can be moved around as needed to be able to input all necessary points.

These input arms are available from several sources and can be obtained with arms that range from about a foot long to about three feet long. They vary in cost from a few thousand dollars to about twenty thousand dollars.

In general, input arms can locate a position within its working envelope to somewhere between ten and twenty thousandths of an inch.

Another method of obtaining actual geometric points is to use the CNC router itself in conjunction with a Hand Held Programmer or Programming Probe to pick up the points from an actual molded part. These points not only take into account variation in the part from design to final part but also variation in the trimming machine. There are, however, several cautions about this approach.

First, a machine generated point is not geometric data. Instead it is an "M" and "G" code program describing the position of each axis of the machine. A type of reverse post processor is needed to translate these axis positions back to a geometric point in space. These programs, designed to back feed machine points are not common, so make certain that software is available for any machine you intend to use in this manner.

Another area to consider concerns overall machine positioning accuracy. We stated earlier that this method of picking up points using the machine does correct for any machine inaccuracy and this, in fact, is true. This correction, however, is only good if the next time the part is run, it is machined on the exact same machine and is located in the exact same location on that machine as when the points were first generated. If the machine has any significant inaccuracy, the resulting trim program will only operate properly on that one machine. If the program is moved to another machine with different positioning accuracy, the resulting trim path may be far enough off that the part is unusable.

Building an accurate five-axis machine is quite involved. There are many machines in the market where the positioning accuracy is such that programs developed using one machine will not execute properly on any other machine, even the same model from the same manufacturer. Needless to say, generally these machines will not be able to properly execute programs developed using off-line CAD/CAM systems and coordinate measuring machines. To utilize this type of machine, it is mandatory that the program be developed using the machine itself and that the program be restricted to operating on only that one machine.

If your tolerances are wide enough and the CNC router accuracy is tight enough, it may be possible to develop trim programs that will be able to execute, unchanged, on two or more machines. In many thermoforming applications, Thermwood five-axis CNC routers routinely share the same program so I know it is possible in real world production. Many customers I have talked to, however, significantly underestimate the difficulty in doing something as simple as moving a program from one machine to another. This area can be quite complex and I have devoted an entire chapter in this book to machine accuracy.

Once you have actual part geometry, there are several options depending on the CAD/CAM system you are using. Some CAD/CAM systems restrict program motions to the three major axes of the machine. They allow the head to be positioned and then perform the actual cut using only the linear axes. These systems are five-axis CAD/CAM systems but are three-axis versions of five-axis systems. The three axis limitation is not necessarily a problem if the machine being programmed is also restricted to three axis simultaneous motions. The limitation to both the machine and the CAD/CAM system, however, results in less efficient programs at best, and may mean that certain parts cannot be trimmed at all in the worst case.

Several CAD/CAM vendors have added a capability to their products to create a true five-axis simultaneous motion using a cut line and head vectors. These systems start with full part geometry. To this is added the trim line. Along the trim line, vectors defining the head orientation are specified. The system then creates a trim program where the tool tip follows the trim line and the head orientation moves smoothly through the defined vectors.

Another similar method of developing a trim path is to create a surface. This is in essence a three-dimensional model of the part. A trim path is then projected onto the surface. The orientation of the head to the part surface is then specified. The program then creates a tool path where the tool tip runs through the projected trim line and the head maintains the specified orientation to the surface of the part.

For example, you could define the tool path and specify that the head remains perpendicular to the part surface being trimmed. The system would then create a trim path where the tool tip traveled along the trim path and the head remained perpendicular to the surface of the part throughout the cut.

Another approach used by some CAD/CAM systems is called the ruled surface trim. This approach does not require complete part geometry. Instead, the trim line must be defined in space. From this line, a second offset line is defined away from the part and away from the trim line. It is assumed that a vector from one line to the other is the vector along which the head lays. The system then creates a program where the tool tip travels along the defined trim line and the head is oriented along the offset line.

Another approach is to define the trim line in space and then, at various points along the line, a vector defining the head orientation is specified. The system then creates a trim path where the tool tip travels along the trim path and the head moves smoothly from one defined orientation to the next. Some companies call this synchronized programming.

Earlier we talked about Thermwood's CADpath program. This system is quite similar to the synchronized programming we just described. With it, points are picked up along the trim path using the probe or hand held. These points define both the trim path and the head vectors. Then, without having to convert these points to geometry and back, the CADpath program generates a ready-to-run trim program. The full five axis trim program moves the tool tip through the trim path while smoothly changing the head orientation through the defined points. Even more amazing is that the CADpath program is a standard part of the Thermwood control and requires no additional investment. Also, the CADpath program requires much less training and technical expertise than is required to operate a full CAD/CAM package.

These are the most popular method currently used to create trim programs using CAD/CAM systems. All methods are not available from all vendors. Also, some vendors sell a very basic system and then add some of the capabilities and features we have just described at an extra cost. As this book is being written, systems that operate using the methods just described cost somewhere between $6,000 and $50,000.

The second major application for CAD.CAM systems is in generating surfaces. These surfaces can be combined together to generate programs to machine patterns or molds. As you might imagine, creating a program to machine a surface is quite a bit more complex than generating trim lines.

There are two reasons for generating a surface program. One is to manufacture a product in volume where a surface must be machined into the product. The second is to create a single example of a design such as a pattern or mold.

If you need to produce a volume of parts with surface machining and if a sample of the surface already exists, Thermwood's probe has a scan mode where it can scan the sample and create a program that can reproduce the surface on a production basis. This same process is extensively used in the furniture industry to program carved wood parts for high-end solid wood furniture.

If, however, you are trying to machine the first article of a new design, a sample does not exist so it is not normally possible to use the probe and scan the surface. In this case you must design the part and develop the program using a CAD/CAM system.

CAD/CAM systems intended to work with three-dimensional parts come in two, three four and five axis versions, even though all may be able to program a full five-axis machine. Generally, the two-axis systems are the lowest cost. As additional axes are added, each step further increases the system cost. To understand this, we will need to talk about the different ways of developing a machined surface.

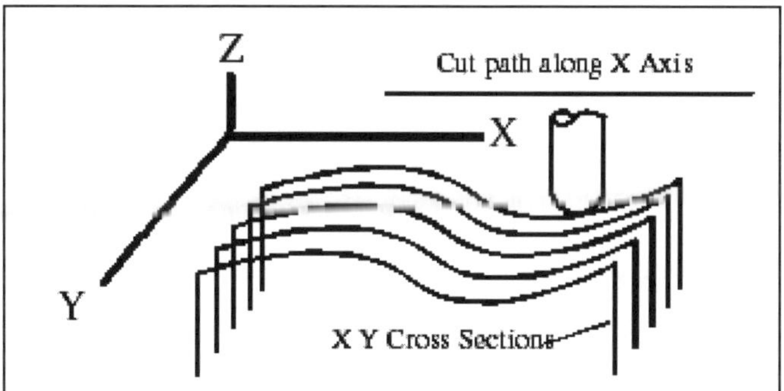

2 Axis System

Let us start with a two-axis system. A typical two-axis system uses cross sections of the part to be machined to determine the path of the head. In a two-axis system, these cross sections must lay along one of the axes of the machine. For example, the cross-section could include the X and Z Axes. To cut this cross-section, the head is oriented and the Y Axis is set. Then, the X and Z Axes move to trace the particular cross section shape. When the cross section is complete, the Y Axis shifts over a set amount and the cross section for that new Y position is again traced using the X and Z Axes.

X and Z cross sections for each point along the Y Axis are cut until the entire surface has been machined. You will note that during this entire process, only the X and Z Axes were used to develop the surface cuts.

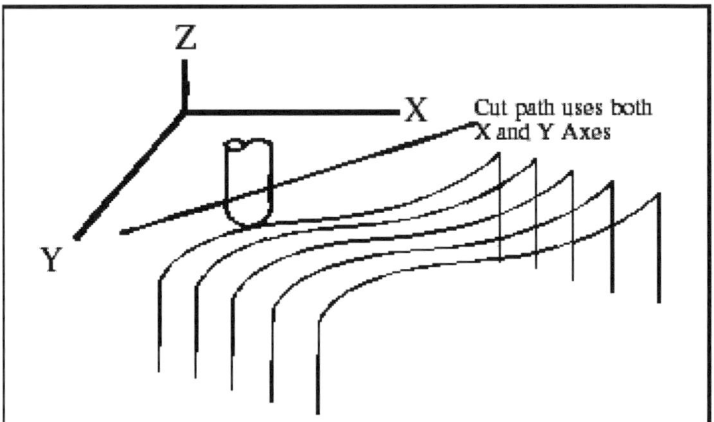

3 Axis System

A three-axis system works in the same way, except the cross section does not need to lay along an axis of the machine. The cuts can be diagonals using both the X and Y Axes. Again, the head steps over each pass but this time the step over is perpendicular to the cross cut direction.

The examples we have covered are what is known as single surface parts. That is the design is made up of only one surface. If multiple surfaces need to be machined using a single surface CAD/CAM package, each surface needs to be programmed and executed separately. The area between the surfaces that blends from one surface to the other must then be finished in some manner.

If you want to be able to design a part with multiple surfaces and machine the blend or transition areas between the surfaces, you need to upgrade to a multiple surface CAD/CAM package.

This package will create smooth transitions between different surfaces after you defined the surface. With most CAD/CAM packages multiple surface capability is an extra cost option or add-on to the basic package.

In each of the packages described so far, we have been working with two or three axes. The actual cuts are performed with the head in a fixed position. The head position can be reoriented between cuts but the head cannot be moved during the actual machining operation. To add the ability to move the head during the machining operation we need to move to a four or five axis CAD/CAM system.

With these more advanced systems come more capabilities. It is now possible to perform what is commonly known as flow line surfaces. When executing flow line surfaces, the system tries to keep the head perpendicular to the surface being machined within a specified range. This can be useful when using shaped or flat bottom cutters to provide smoother surface finishes. For an additional cost, lead-lag can be added to the package. Lead-lag capability means that the head is not exactly perpendicular to the surface, but instead leads or lags the surface by a fixed angle.

This can be useful when performing machining functions where the tool must be angled slightly to provide some type of machining clearance. Cutting aluminum honeycomb is a good example of this.

Honeycomb is machined using a cutter that looks very much like a gasoline engine valve. For this reason it is called a valve stem cutter. This cutter must be tilted a few degrees into the part so that the trailing edge is slightly above the surface that was cut with the leading edge. Not only does this keep the trailing edge from bending or turning down the edges of the honeycomb but it helps to lift the cut material from the part. Lead-lag capability is quite important in this type of application.

Again with full five-axis systems you can have single surface of multiple surface capability. Another upgrade for multiple surface capability is multiple surface flow line capability. This means that not only does the system generate the transitions between surfaces but allow flow line machining of the transition areas. A further enhancement is gouge checking.

With gouge checking the system checks the tool and head to make certain that in machining one surface, trying to stay perpendicular, it does not hit or gouge out another surface or area on the part. Without gouge checking, a problem of this type results in either a gouged out part or a head crash.

When a lower cost system with gouge checking detects interference it alerts the operator or programmer. It is then the responsibility of the operator or programmer to modify the program to eliminate the problem. High end systems can modify the programs automatically by changing the angle of attack to eliminate the interference problem.

Another feature of some CAD/CAM systems is called trimmed surfaces. When using trimmed surfaces, a part with two or more surfaces allows the programmer to create oversized, overlapping surfaces and then trim them to fit together. This can be quite a bit easier than attempting to develop exactly sized surfaces and then fitting them together.

Another method of creating a surface is called NURBS. NURBS stands for Non Uniform Rationalized B-Spline. This is a very complex name for a very simple idea.

Using NERBS, a surface is defined by defining the paths that surround the surface. For example, four curved lines could be connected at the corners. This system then creates a surface that flows to connect the four lines smoothly. Although there is a lot of very high-end math involved in the creation of the surface, NURBS offers the programmer an easy way to develop a smooth flowing transition between known lines.

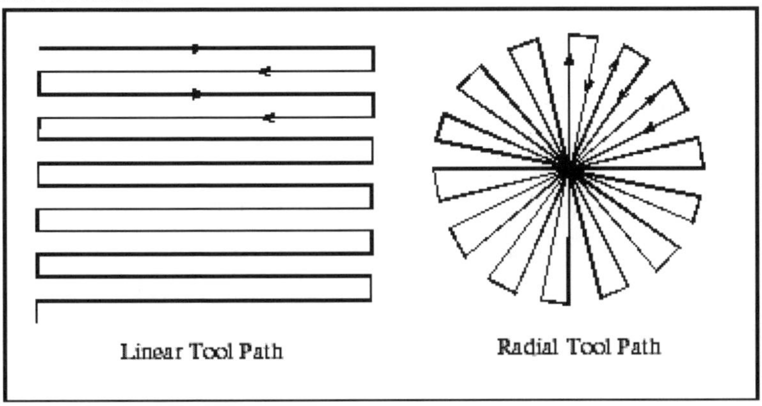

Each of the systems we have discussed so far have used linear tool paths. Some companies call this parallel machining. It simply means that the cut paths are straight lines parallel to each other, each offset from the last. There is another method of machining called radial tool path that may work better on certain parts. In radial tool path, the cutting path is a straight line from a center point to an outer radius and then back to the center. Each time the path starts from the center point but goes out on a slightly different angle like the spokes of a wheel. It continues shifting the angle until it goes out at every radial of a complete circle.

The choice between linear tool paths or radial tool paths depends on the shape and surfaces of the part. In some cases one method or the other will give you significantly smoother surfaces to work with by reducing steps on nearly vertical surfaces. This results in significantly less post finishing of the part.

In either case, however, unless the steps are made extremely small, some sanding or post finishing will be needed. If you use very small steps, the cycle times can become extremely long, in most cases, costing more in machine time than the required post finishing. The balance between more machining time and finishing time will vary with each company and each application.

So far in our discussion of CAD/CAM programming, we have been talking about programming the finished surface of the part. There is another area of programming that is almost as important. This area is roughing.

Roughing or broaching is the process of removing the bulk of material from a part blank so that the finish program is only required to remove a small amount of material to achieve the final, smooth surface.

These roughing programs tend to remove material in much coarser steps than the finish program. Almost every CAD/CAM system uses a separate set of menus and programs to create the roughing program. On higher performance systems, it is only necessary to define the maximum cut and the offset of the roughing program from the finish surface. The system then develops the roughing program to cut away the bulk of material from the blank. The result of roughing is a part that is slightly larger than the finished piece. The amount that the rough part is oversize varies depending on a number of factors. Typically the rough part is at least .030 to .060 inch oversize and can be a half an inch or more oversized in some materials. To achieve a smooth final product, however, the roughed blank should not be so large that significant cutting forces are developed during the finish cuts. These high cutting forces can result in degraded surface quality and "push-off" where the tool and head flex in reaction to the cutting force. The balance between rough cut and finish cut comes with experience.

Because of the limits of a book of this type, I have only offered an overview of programming. What I have explained are the fundamentals. Advances are occurring every year and new, easier, faster programming systems keep coming. Advances in the last ten years have been nothing short of incredible. CAD systems that cost millions of dollars can now be purchased for under $20,000 and the new systems work better. Programs that used to take weeks to perfect can now be made in minutes using a programming probe. It is likely that the fundamentals I have presented won't change much. What will change is that simpler and faster methods will become available.

One final caution about advancing technology. Make certain that the company you purchase your CAD/CAM system and CNC router from have a policy of offering system upgrades rather than requiring you to scrap your investment and buy a new machine or package to get new features and technology.

One thing you can be sure of is that these technologies will continue to advance. Don't lock yourself into a system where you can't continuously advance with advancing technology. These systems should be purchased to make your company competitive. To remain competitive you can't allow your competition to buy capabilities that you can't easily add to the systems you rely on every day.

Chapter 6
Accuracy

Accuracy

In twenty-five years of dealing with CNC routers, accuracy has been the single most misunderstood aspect of the product. Virtually nobody I met from the wood or plastics industry has had a reasonable understanding and less than one third of the engineers from the aerospace industry possessed an accurate understanding.

In most practical applications, this misunderstanding has not had a significant impact. The capability of the machines has been much better than required for the specific application so acceptable parts have been produced despite the fact that those producing the parts didn't understand why.

Some unscrupulous machine manufactures, however, have used this general lack of understanding to try and promote their machine as being substantially more accurate than their competitors. Over the years we have concluded that some of these companies actually believed their claims.

The root of this general misunderstanding is that people make a series of unconscious assumptions and thus substantially simplify their understanding of machine accuracy.

If a machine is accurate to within plus or minus five thousandths, it should be able to machine any part that will fit on its table within plus or minus five thousandths, right?

The answer to this question is absolutely no. A general rule of thumb is that most people can machine parts to a tolerance two to five times greater than the positioning accuracy of the machine. A machine that can position within five thousandths can generally produce parts consistently within ten to twenty-five thousandths, but that isn't even totally correct. Before we add further to the confusion, let us back up and look at the fundamentals.

Accuracy exists in layers. The top, simple layer is the one most people partially understand. It is the largest layer and has the greatest impact on machine accuracy. As we attempt to achieve tighter and tighter accuracy, we enter the lower and lower layers. In these layers the physics of the world changes. We are dealing with smaller and smaller numbers and physical effects that are not even measured in the top layer begin to dominate as we try and achieve higher and higher accuracies.

Most people get in trouble by attempting to achieve accuracies several layers deep with only a partial understanding of the top layer. In the remainder of this chapter I will begin at the top layer and move through subsequent layers as we try and achieve tighter and tighter accuracy.

First, what is accuracy? There is even a misunderstanding of accuracy. Let us try an experiment. Place a dial indicator on a machine and run an axis against it. Zero the indicator. Now run the axis out and return. If the indicator is within a thousandth of an inch, does that mean that the accuracy of the machine is a thousandth of an inch. No it doesn't.

This test does not measure accuracy, it checks single axis repeatability. This is an extremely limited measurement and is only useful as part of a more comprehensive measurement effort. Despite this fact, I have seen companies use this very test to prove just how accurate their machines are. Let me give you an example.

In this test we are able to repeat within a thousandth of an inch but the actual position of the axis could be an inch or two different than the position the control intended. The machine, which you thought had a thousandth of an inch accuracy is actually inaccurate by up to two inches.

This test also only measured repeatability along a single axis. The axis perpendicular to the one we measured could have been a half an inch different than its original position. While we were assuming a repeatability of a thousandth of an inch, the actual repeatability of the two axes together was about a half an inch or five hundred times worse.

This axis could have fifty thousandths backlash. If we approach the point from one direction we repeat within a thousandth of an inch. If we approach the exact same point from the opposite direction we are off by fifty thousandths. Officially the repeatability is still a thousandth of an inch but that number leaves an overall impression about the machine that is not totally accurate.

Before we continue, we need some definitions. Even this isn't very simple. Many years ago I was part of the trade association that concentrated on industrial robots. They decided that in order to clear up the confusion surrounding the accuracy of industrial robots, they would adopt standard definitions for accuracy and repeatability. The ensuing confusion lasted almost a year. I left the group before they developed their final definitions but I assume that they finally were able to agree on specific language.

For our purposes, I will offer some relatively simple definitions that we will then use in the remainder of this chapter.

First, accuracy. When you say the word accuracy to most people they assume you mean multi-axis absolute positioning accuracy. Absolute positioning accuracy means the ability to achieve a point in space within a certain tolerance. Normally this is a distance measured from the reference, 0,0 or home position of the machine. If the control tells the head to move 35 inches, absolute accuracy tells us how close to thirty-five inches did the head actually move. This, however is single axis absolute accuracy since only one axis was involved in reaching the thirty-five inch position.

If we moved the X Axis, we made some assumptions. We assumed that since we didn't move the Y or Z axis, they are both still at the zero position. Most likely this is not the case if the axes are interdependent at all. Interdependent means that when one axis moves it carries with it another axis. If the Z axis is mounted on the Y axis, moving the Y axis will also move the Z axis. The position of the head on the Z axis doesn't change but the entire axis has been moved. In moving the axis it is very likely that its orientation to the machine has changed slightly. This slight change will have moved the head in relation to all the other axes.

Although we have been able to position along the X axis within a thousandth of an inch, we may find that the Z axis has shifted left or right more than a thousandth of an inch. The single axis absolute accuracy may be a thousandth of an inch but the multi-axis absolute accuracy is likely a substantially larger number.

Repeatability is the ability to return to a point in space each time a program is executed. Just like absolute accuracy, repeatability can be either single axis or multi-axis. On all systems, repeatability is always much better than absolute positioning accuracy.

Another type of accuracy that can be important in CNC router applications is relative accuracy. This is a measure of how accurately you can position from one location on a part to another. For example, how accurately can you locate the center of two holes with respect to each other.

The accuracy and repeatability we have discussed so far are positioning accuracy, unloaded in free space. This means that the head can position to these tolerances as long as there are no external forces exerted on the head. A machine with an absolute positioning accuracy of five thousandths of an inch might shift twenty-five thousandths when the cutter enters the material. This machine still has an absolute positioning accuracy of five thousandths, but its practical operating accuracy is much less.

This situation becomes even more complex when we consider that the machine will encounter different levels of force depending on the material being cut, the amount of material removal, the speed and several other factors. Also, the machine will react differently depending on head orientation and the direction from which force is applied. Some machines may also react differently in different parts of the working envelope. This means that trying to determine a practical operating accuracy is all but impossible.

Our discussions so far have been at the top level of accuracy. We are assuming a fixed, rigid world that is physically unchanging. Let us continue at this level and consider the factors that affect the ability of a machine to physically achieve a theoretical point in space.

Let us start with the three linear axes. To achieve a specific point in space, the first requirement is that when we tell an axis to move a specific distance it actually moves that distance. We have already discussed lead screws and lead screw compensation. A good tight, accurate lead screw compensation system will address this area. The only caution here is that many commercial CNC controls really skimp on the number of compensation points available. As you try and achieve tighter and tighter tolerances, it is necessary to compensate at closer and closer intervals. If the control restricts the number of compensation points, it may be difficult to achieve the positioning accuracy desired.

Another factor that affects the ability to accurately position an axis is the rotary resolution of the servomotor. The linear position of the axis is determined by the rotary position of the servomotor that drives the axis. A rotary encoder controls the rotary position of the servomotor. The resolution of this encoder is a measure of how small an angle of rotation the encoder can measure. Encoders today tend to be optical and so a complete rotation of the motor is divided into a specific number of steps. The resolution of the axis is the linear distance moved for each step in the rotary encoder.

If we have a resolution of five ten-thousandths of an inch, does this mean we can position to five ten-thousandths of an inch. Again the answer is no.

The control requires at least two steps to achieve a stable position. It generally requires at least one more step in each direction to even know that a position error exists. We already have a band width of four steps and that is for a theoretically perfect system. In actual practice, a high quality, tight, properly adjusted servo system can normally position within plus or minus ten times the encoder resolution. Thus, the system we just described with five ten-thousandths resolution can be expected to achieve plus or minus five thousandths positioning accuracy.

I have seen CNC router companies claim positioning accuracy equal to their encoder resolution. In these cases, these companies were claiming accuracy that was at least ten times better than is technically possible using the traditional definition of single axis positioning accuracy.

So why don't these companies simply use higher resolution encoders to achieve even higher accuracy?

The problem that occurs is that the control must count every pulse from every encoder as it occurs. If resolution is one ten-thousandth of an inch and the axis is moving at 1,000 inches a minute, the control must be able to count one million pulses per second on each axis. This is about the limit today and many controls can only count a third to half this many pulses per second. When this occurs, the choice becomes either fewer encoder steps and a wider tolerance band or slower feed speeds.

Currently high-end controls like Thermwood's SuperControl can handle over a million pulses per second. This allows encoder resolutions of one ten-thousandth of an inch and still supports high feed speeds. Many metalworking CNC controls, however, have much lower feed speed requirements. Because these requirements don't need high pulse count capability, these controls can present very real limitations when they are used for control of high feed speed CNC routers.

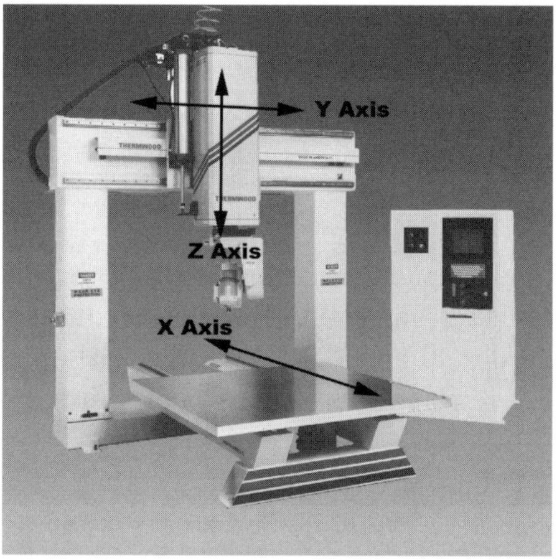

Once we are sure that the axis will move the proper distance, we must make certain that the track on which it is running is relatively straight. Errors in rail straightness have little effect on the axis being moved. Instead, these errors manifest themselves in errors caused by shifting an axis that is perpendicular to the axis being moved. If you are moving the X Axis and it bows three thousandths up, the X Axis position will not be affected. The Z Axis position, however, will have shifted by three thousandths. If the rail bows both up and to the right, both the Z and the Y Axis will have shifted.

This effect can be multiplied when axes are interdependent. This occurs when axes are stacked on each other. Let us take a typical moving table five-axis machine as shown on page 206. The table or X Axis, which moves front to back is not interdependent with any other axis. If the rails supporting the table are bowed three thousandths upward, it will cause a simple three thousandths effect on the location of the Z Axis since it is the relationship of the table axis to the Z Axis that determines the Z Axis height.

The Y Axis, however, moves back and forth on rails mounted to the top and bottom of the bridge. It also carries the Z Axis and the Z Axis carries the two rotary head axes. These four axes are interdependent since errors in one axis tend to add to errors in subsequent axes.

Errors in interdependent axes also tend to multiply. Let me show you one example of this effect.

Let us assume that the rails on the machine bridge are each bowed a couple of thousandths of an inch. Achieving rail straightness and parallelism of a few thousandths of an inch is no trivial matter. In aerospace applications it can take weeks of adjustment with laser interferometers and cost tens of thousands of dollars to align linear axis rails to this level.

If the top rail is two thousandths of an inch out and the bottom rail is two thousandths of an inch in, the two are four thousandths of an inch different. With a typical 16 inch spacing between rails, a three foot Z Axis and a 14 inch pivot point on the rotation of the spindle motor, a two thousandths error on the rails will cause a twelve and a half thousandths error at the tip of the tool. If we wanted to reduce that error to five thousandths, we would need to align the rails to less than one thousandth of an inch. At this point, we will begin to experience effects from level two and level three factors. The only other alternative is to reduce the stroke of the Z Axis and the pivot point of the router spindle. All this, and we have only examined the effect of the first of four interdependent axes.

We must now consider the rail straightness of the Z Axis slide. Again, there is s multiplier effect. The bearings that slide on the Z Axis assembly have a bearing separation that is not much over one third of the stroke of the axis. Any bow in the rails will then be multiplied by about three. A two-thousandths error will result in up to a six thousandths positioning error at the end of the tool. We are now up to eighteen thousandths off even though we have achieved rather tight tolerances for rail straightness. We still have two axes to go.

Alignment of the machine can also have a significant effect on positioning accuracy. The first alignment requirement is that the linear axes are exactly square and perpendicular to each other. In determining the perpendicularity of two axes, any bow in the axis must be accounted for so that the average path of an axis is perpendicular to the average path of the second axis. If the end two feet of an axis were bowed slightly and this section was used to establish perpendicularity, the end of the axis could possible be rather far off. Also, if the end of the axis were used to establish perpendicularity and that end were off, the average path might be rather far off. The third axis must then be perpendicular to both the other axes and the complexity becomes a order of magnitude more difficult.

Accurate alignment of linear axes on a CNC router is a science, but luckily, this effort is normally completed at the factory by factory trained technicians. Except in extreme circumstances, once the linear axes of a machine are aligned, they tend to stay in alignment.

Now we move to the two rotary axes, axis 4 and 5. Aligning the linear axes of a machine is child's play compared to achieving accurate alignment of the rotary axes of a five-axis machine. There are many factors associated with properly aligning a five-axis head. We will look at many of these factors.

Thermwood has developed a ten point head alignment procedure that can achieve simultaneous alignment of all of the factors. Developing this procedure alone was a major technical breakthrough. When aligning the head axes, an adjustment in almost any parameter will affect the adjustment of other parameters. Prior to this procedure, it was common for a technician to "chase his tail" trying to align a five-axis head. Each adjustment threw something else off. Adjusting that knocked something else off. We went around and around until either we got "close enough" or we gave up.

Thermwood's head alignment procedure achieves alignment in a very precise order. That order minimizes the effect I just described. There are several points about this procedure that should be kept in mind.

First, the procedure must be started at step one and must proceed, one step at a time to step ten. If any step is skipped or performed out of order it won't work. If any single step is performed without going through the entire procedure, it won't work.

I am not going to go through the entire procedure in this book. All Thermwood five-axis machine manuals detail this procedure and both text and video explanations of the alignment procedure are incorporated into every Thermwood five-axis control system.

I am, however, going to discuss the various areas in the head where alignment is important. I will also discuss machine design parameters that can affect accuracy in each of these areas.

First, why do we need to align the head? If the head was properly aligned at the factory, why should we need to realign it?

This brings up a major controversy. Most five-axis head assemblies are designed with alignment plate that allows the head to be adjusted to achieve proper alignment. The question is, just how rigidly do we attach these plates together.

If we simply assume normal operation, the answer is simple. The plates should be secured so that they cannot move under any circumstances. This way the head will never go out of alignment.

The problem is that we can't count on this normal operation. Through operator error, it is all too common for the machine head to crash into the table or a fixture. The servomotors that drive the axes of the machine have current sensors that can detect the high current that results from one of these crashes. The sensor, however, is not instantaneous, so rather large forces can develop before the system errors out. Also, if the entire Z Axis assembly is moving at high speed when the crash occurs, a large amount of kinetic energy from the rapidly moving mass must be dissipated.

If we rigidly secure the adjustment plates, these crash forces are directed into the head where they will permanently damage the gears, bearings, castings and other expensive components.

To prevent this damage, the bolts connecting the adjusting plates have been designed so that they can slip in a crash. This slip is supposed to occur just before internal components begin to fail. Friction of the adjusting plates can absorb substantial energy in a crash.

The idea here is that when a crash occurs, normally all that is required is that the head be realigned. A problem occurs, however, when these machines are used for heavy cuts. The head can be moved out of alignment even though a crash didn't occur.

Each five-axis manufacturer must handle this in their own way. Thermwood has been approaching the problem from two directions. First, they have been trying to beef up the internal components of the head so that they are more resistant to crash forces. For example, keys that secure components to shafts within the head have been increased in size and have been switched to a heat-treated, hardened material from soft steel. Prior to the switch, keys could actually be smashed in a crash, opening up a small amount of lash. When the head was realigned, it then required very little force to move the key slots and knock the head out of alignment again.

With redesigned internal components, Thermwood has added additional fasteners to substantially increase friction between adjusting plates. This will make it much more difficult to knock the machine out of adjustment, but, it is at the risk of more serious damage in a major crash.

Experienced five-axis programmers and operators seldom crash the machine. Inexperienced operators and programmers crash the machine often. Today, experienced customers are concerned with going out of tram too easily. With these changes and higher maximum feed speeds, the changes may result in more serious repairs when a problem does occur.

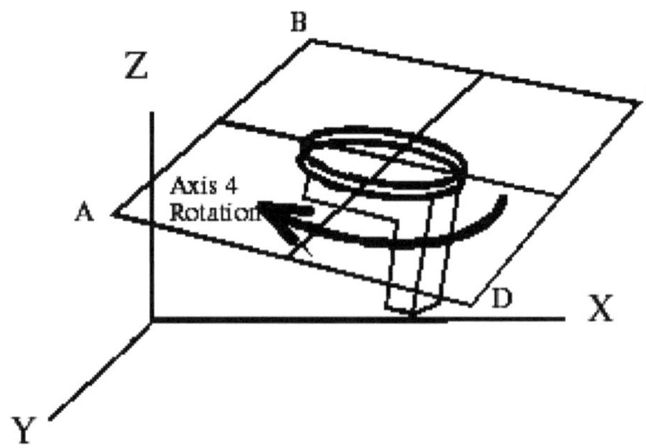

The first requirement for head alignment is that the plane on which Axis 4 rotates (A-B-C-D above) must be exactly parallel with the plane formed by the X-Y axes, and exactly perpendicular with the Z Axis.

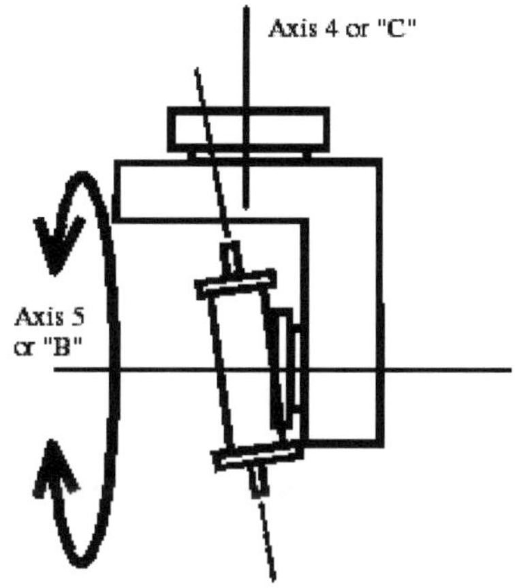

Next, we must make certain that Axis 5 is exactly perpendicular to Axis 4. We must also make certain that a line through the center of the router spindle is exactly perpendicular to Axis 5. The center of rotation of Axis 5 must also pass directly through this centerline of the router spindle. It cannot be shifted forward or backward

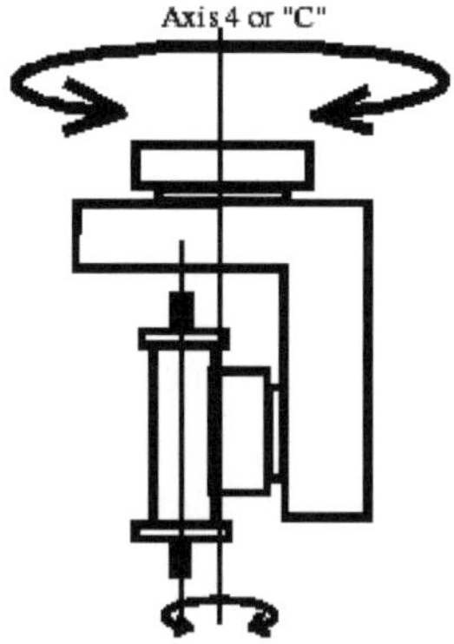

If the center of rotation of Axis 5 does not pass through the centerline of the router spindle, the tool will spins in an offset circle when Axis 5 rotates. This will make it difficult or impossible to accurately machine any path that uses Axis 5.

Once these basic mechanical connections have been adjusted, it is necessary to adjust the actual rotation of each axis.

The zero reference or home position of each of the two rotary axes must be adjusted. Axis 4 must be adjusted so that in the home position, the Axis 5 centerline is exactly parallel to the X Axis and exactly perpendicular to the Y Axis.

The home position of Axis 5 must also be adjusted so that when the axis is instructed to point straight down, it is, in fact, exactly pointed straight down.

Just as there was run out in the ball screw drives of the linear axes, there is also inaccuracy in the worm and spur gears that drive the rotary axes. If you recall, we used lead screw compensation to correct for any errors in the linear axes. In the same manner, we must also use lead screw compensation to correct for pitch errors in the rotary axes.

There are several sources for errors in the worm and gear drives. The first area for problems comes from a worm gear that lopes. This means that the gear and the gear shaft are not exactly concentric. When this occurs, the gear runs tight at the high point and loose at the low point. The result is that lash exists at some point during each revolution of the worm gear. There is no way to correct this problem using compensation, and in general, the only solution is replacement of the worm gear.

For this reason, a good worm gear design for a five-axis head is a single piece design where the shaft is turned and ground and then the actual gear shaft is used to cut the worm gear itself.

Just because the worm gear is cut perfectly doesn't mean that it won't lope. Any major crash places high loads on the worm gear. It is possible to actually bend the gear shaft slightly what has the same effect as a poorly machined gear.

The next area of concern is the large spur gear that the worm drives. In order for the worm gear to turn, there must be a very slight clearance between the back of the worm gear and the next tooth in the spur gear. This clearance is called backlash and it is necessary to prevent the worm and spur gears from binding.

If this backlash becomes too large, however, it allows the axis to move back and forth causing a loss of accuracy. In this area we again run into a multiplier effect, but the effect will depend somewhat on the orientation of the head. If the head is pointed straight down, backlash in Axis 4 will have absolutely no effect. If, on the other hand, the head is horizontal, a one thousandth backlash will translate to three or four-thousandths at the tool tip.

Another factor tends to make this problem even worse. When the spur gear is cut, it is all but impossible to make the thickness of every tooth exactly the same. It is common for the teeth to vary slightly from one to the next. The gear mesh must be adjusted to accommodate the widest teeth so the axis doesn't bind. This means that backlash is even greater when the worm is engaged with a narrow tooth.

One way of helping this situation is to perform a wear-in procedure during manufacture. The worm and spur are meshed so that the worm actually rubs both teeth at the thickest point. It is then run until the thicker teeth wear slightly. Then the mesh is tightened again and the procedure repeated. The gear mesh is constantly moved in, a little at a time until a reasonable match is achieved and backlash is both small and consistent.

Normally in this type of system the worm gear is made of a hard material such as hardened steel and the spur gear is made of a softer material such as bronze. This allows the wear-in process to work.

Once a backlash clearance is established, the bronze gear does not continue to wear quickly. It will, however, wear over time with normal use. This brings up another problem.

Most part programs cut with the head pointed more or less downward. It is this area of the gear that will wear over the years. The head must, however, be able to rotate up to the zero or home position. This means that the additional lash in the downward working position cannot be adjusted out without causing the gears to bind toward the home position. Older machines will, therefore, have to deal with additional inaccuracy due to increased gear backlash.

Because the gear making process is not perfect, not only the thickness of the gear teeth vary, but the amount that the gear rotates for each revolution of the worm gear will also vary.

When the control instructs the axis to rotate one degree, the actual rotation may or may not be exactly one degree.

In the past, Thermwood measured exact positions each 45 degrees or rotation on each axis and input compensation at these points. We then varied the compensation linearly between these measured points. Recently a series of precise tests has shown us that even high quality gears can exhibit variations in as little as a few degrees of rotation. Because of this, we have started compensating the rotary axes every one degree to minimize the effect of these variations.

This compensation is important for rotary axes because of the relatively long pivot point of the axes. With a 14 inch pivot point from the point of rotation to the tool tip, a one one-hundredth of a degree error translates to almost two and a half thousandths of an inch.

At this point, we have looked at several of the level one factors that affect accuracy that are associated with the mechanics of the machine. There are other major factors that can also influence accuracy.

In Chapter 2 we talked about dynamic path following of a servo system. The ability of a particular servo system to dynamically track a path is a major factor in part accuracy. Even with an essentially perfect machine, if the servo system is not extremely tight, path following and the resulting trim path will not be accurate.

The acceleration and deceleration rate of a machine can also have an effect on accuracy. In general, accelerating a large five-axis machine is more difficult that accelerating a lighter, less massive three axis machine. Vibration, flex and push-off can occur during the acceleration and deceleration of a motion.

Most five-axis machines have their acceleration set at a compromise between fast cycle times and smooth, accurate path following. In some instances, a smooth accurate path is much more important than speed. In these cases, at least on the Thermwood SuperControl, the acceleration and deceleration rates can be reduced in the program, to allow the motions to be accurately made. This is a useful feature since it is only necessary to reduce the rates for very specific motions. Once the moves are complete, the rates can be returned to factory settings. This means you can get the short term advantages of slower acceleration and deceleration rates without having to slow execution of the entire program.

Another factor that affects accuracy is the length of the tool. Every program assumes a specific length for the cutting tool. If the bottom of the tool is used in the machining process, the depth with which the tool is inserted into the collet or tool holder will directly affect part tolerance.

Thus, if the tool is inserted five or ten-thousandths of an inch high or low, the part will be five or ten-thousandths of an inch off due only to the tool length.

Router bits are held in place by a collet. The collet is a round tapered holder that fits into a tapered spindle shaft or tool holder. The collet is then tightened by turning a nut that forces it into the taper of the spindle. The more you tighten the nut, the farther the collet wedges into the taper and the tighter it holds the tool. Herein lies the problem.

By varying the amount you tighten the nut you can vary the distance the collet is forced into the taper. This simple fact makes it very difficult to achieve an accurate tool length. As the nut is tightened, it not only pushes the collet up but also pushes up the tool bit. Sometimes the collet is secured in a tool holder that has its own taper. This tool holder then fits a matching taper in a tool change spindle.

When a router bit is fastened into a tool holder, normally a reference ring on the tool holder itself is used to measure to the tool tip. The same difficulty we just described will occur, but even if the tool length can be adjusted exactly right, all tool holders may not fit into the tool change spindle exactly the same. Not only do we need to worry about the tolerance between the tool and the collet, but we must also worry about the tolerance of the tool holder itself.

Because of these problems, it is virtually impossible to achieve any level of tool length accuracy by mechanical means. If the tool tip is to be used in machining or tool length is important to a process, another approach is required.

To address this problem, Thermwood developed five-axis tool length compensation for its SuperControl. This feature is also available on aerospace versions of several commercial control systems although it is normally not included on CNC routers that use these systems.

With this system, a sensor, located off the back of the table, is used to measure the exact length of the cutting tool. The sensor is actually a metal plate that is given an electric potential. The tool is moved to a position just over the plate and lowered in one tenth of an inch increments until it contacts the plate. At that point, the control knows that the tool has contacted the sensor because the electrically charged plate has been grounded to the machine through the tool. The Z Axis then moves up in hundredth of an inch increments until the ground signal disappears, indicating that the tool is no longer touching the plate. The head then moves down in thousandth of an inch increments until it again touches. It then moves back up in ten-thousandth of an inch increments until is again off the plate. At this point, the exact axis position is measured and the length of the tool can be calculated very accurately. This entire process takes only a couple of seconds and provides a level of accuracy that is totally impossible any other way.

The new tool length is then used to recalculate the entire part program, modifying it so that tool length will no longer contribute to inaccuracies.

It is possible to get by without the automatic sensor by using an accurate dial indicator to measure the difference between the original tool length and the current tool length and then enter that number into the tool length compensation table. I know of no way, however, to get by without tool length compensation capability if the tool length is critical in your application.

If you plan to use a five-axis machine for model or mold making, it is important that some type of tool length compensation be available.

Another source for inaccuracies in a part is the CAD/CAM system itself. This is tied into the fundamental way a CAD/CAM system creates curved lines. The system does not program a curve, but instead develops a series of short lines that approximate the curve. To develop these lines, the programmer must specify the maximum distance that any point on the line can vary from the exact theoretical curve.

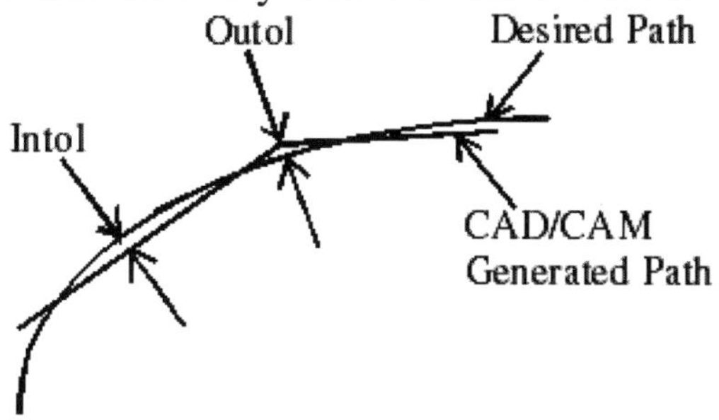

The amount that the CAD/CAM generated line can vary from the theoretical curve is called the in-tolerance and out-tolerance. The higher this number, the longer the segment lines but the more inaccurate the curve.

Small intol and outol numbers result in accurate curve following but there is a price to pay. The large number of small line segments can take significantly longer to process. I have seen complex CAD programs require twenty to forty hours to process. Also, the resulting program can be very large and some control systems may not be able to handle the large file size. Even if the control can handle the large files, block processing speed limitations may result in very slow execution. In either case, many companies opt for larger tolerance numbers. This does, however, result in rougher, less accurate programs.

Therefore, because of the path following tolerance specified for the CAD/CAM system, some variation from the actual desired path will occur. Add to this the fact that a certain dynamic path following error may result from the machine servo systems themselves as they try to execute the short line segments, and you have yet another source for error.

I have seen many instances where a path is generated to trim a three dimensional plastic part and the machine does an inconsistent job even though it claims to have a rather tight repeatability. In these cases, inconsistency means that the trim position on one part differs from the trim position on another.

The key to this problem is that the trim quality is judged by the relationship between the trim line and the actual part itself. In almost every case, the trim path followed by the machine was accurate. The part that was trimmed, however, was not.

It is common in plastic molding for overall part size to vary. This can be caused by numerous factors including variations in molding temperature, raw material formulation, cooling cycle etc.

When an inconsistent part is processed by an accurate machine, the result is inconsistent. Something as simple as accurately locating each part on the holding fixture can result in inaccurate trims.

If the trim accuracy required is tighter than the variation in overall part size, we are faced with a new problem. The only way I am aware of to address this problem is to somehow adjust or vary the trim program to match each part trimmed.

One way to do this is to use a touch sensor to measure certain dimensions on a part and then either modify the trim program using variables and the Advanced Function

Language or use the measurements to select from several different programs, each with a slightly different trim path.

Both of these approaches are technically possible today. They are, however, rather sophisticated and will require a competent computer programmer to successfully execute.

Research is currently being conducted on methods to actually measure a part as it is being trimmed and modify the trim path in real time so that it matches the part exactly. This is even more sophisticated and it is difficult to determine just when such technology might become available.

We have put quite some time into discussing accuracy, yet at this point, we have only looked at the first level. Many of the areas that we have discussed are generally understood by machine users. All of these areas have made the assumption that the machine, tooling and parts are fixed, rigid and unchanging. In truth, nothing is perfectly rigid.

All materials bend and flex under load. Some materials flex more than others and some structures resist flexing more than other structures. When we are trying to achieve a tolerance of a few thousandths of an inch over structures that may span five or ten feet, flexing and movement of the structure become a major factor. In general, it is virtually impossible to build a practical structure that will resist flexing enough to achieve tight tolerances on a five-axis machine.

All the discussion on rail straightness is no longer valid since a perfectly straight rail will not remain straight as the Z Axis carriage moves over it. Even a couple thousandths of flex at the rails will result in five or ten thousandth variation at the tool tip. Add to this the fact that each axis flexes independently and you have an extremely difficult situation.

One approach we have tried to correct this situation is to machine the rail supports with a very slight crown so that the rail becomes perfectly straight when the weight of the Z Axis carriage moves over it. Although this sounds like a viable approach, it doesn't work very well in practice.

The reason it didn't work is that different bridge structures, even though they are identical in design, don't necessarily flex exactly the same. Variations in the steel itself, the welding process and other unknown factors result in structures that are close to the same, but not exactly the same. The accuracy you are trying to achieve is so small that even tiny variations in the structure make it unworkable.

The method we finally turned to, although more labor intensive, does result in a reasonably straight path for the tool head. We install a laser measurement sensor at the tool tip, then run the axis back and forth to determine the straightness of the tip. We then shim the rails until the tip follows a straight accurate path.

This approach does substantially improve the straightness on the axis despite normal structure flexing, however, it is far from perfect.

First, it only sets that path as straight at a certain Z-axis height. Lifting the Z Axis may change the loads on the bridge structure resulting in a different flex and thus a different path.

The path will also vary from dynamic loading. The flexing we have been discussing is caused by the force of gravity on the moving structure. As this moving structure is accelerated and decelerated, new forces are added that also cause structure flexing. The faster the structure is accelerated and decelerated, the greater these forces.

Acceleration flexing is not consistent. When the machine is changing speed these forces exist. When the machine is either stopped or moving at a constant speed the forces do not exist.

The only way to reduce inaccuracy caused by this type of flexing is to reduce the rate of acceleration and deceleration. These two factors, however, greatly influence production cycle time. Decreasing the acceleration rate will substantially increase the amount of time required to move through a programmed path. In general, slower acceleration and deceleration will result in slightly more accurate path following at the expense of longer cycle times.

Some controls, such as Thermwood's SuperControl allow the acceleration and deceleration rates to be changed within a program. Advanced Function Language commands exist that can reset the maximum acceleration and deceleration rates. This means that the acceleration rates can be lowered during a program only for those areas where increased accuracy might be required and can be increased for areas where extremely tight tolerances are not required.

This approach allows for both improved accuracy and fast programs on difficult parts.

Another type of flexing that will affect accuracy is what we call “push off”. As the cutting tool rotates into the material being cut, the tool is pushed away from the material by the force of the cutting process.

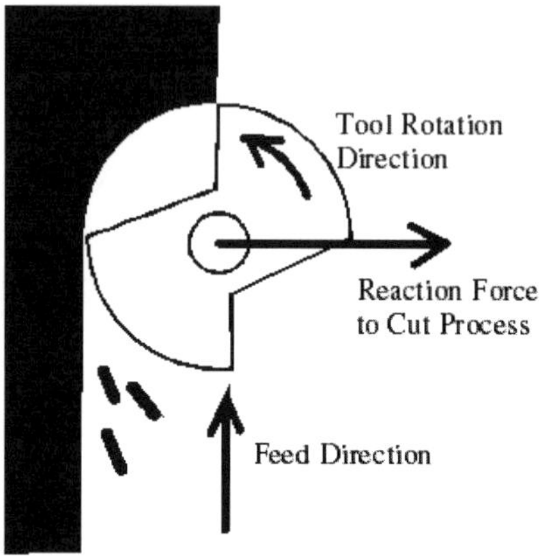

The reaction force to the cutting process shown above pushes the tool away from the part as it is cut. This has several effects on accuracy.

If the router bit is either small diameter, long or both, the most pronounced effect is a flexing of the tool itself. The tool actually bends and runs slightly curved. With smaller tools, this bending or “push off” can be anywhere from a few thousandths to ten-thousandths or more. The amount of tool push-off can be seen by running across a part and then stopping quickly. In the stopped position, the reaction force no longer exists so the rotating tool will relax into the material causing a notch. The depth of the notch indicates the amount of push-off from the cutting process.

Depending on the reaction force, it can also flex the machine itself, further increasing the amount of push-off.

So we now see that achieving absolute accuracy is quite difficult when faced with mechanical alignments, small mechanical inaccuracy and structural flexing. We still are not done.

In all the above considerations we again made another assumption that is not accurate. We assumed that the actual mechanical sizes of various machine components are constant. When we are operating in the range of a few thousandths of an inch, this is not true. In fact, the size of everything changes unless you are able to keep the temperature of everything exactly the same all the time.

Most materials expand when heated and contract when cooled. The amount of expansion and contraction is so slight that it is generally ignored in everyday life. This expansion and contraction can induce significant variation when you are trying to achieve tight tolerances on large machines.

Steel has a coefficient of expansion of 8.4 to 15 millionths of an inch per inch per degree Fahrenheit. This means that a one-inch piece of steel will expand by up to fifteen millionths of an inch for each one degree increase in temperature. This seems like a rather small number until you look at it using real world numbers. Let us assume a temperature difference of thirty degrees, a change from 65 degrees in the winter to 95 degree in the summer. This temperature will cause a typical twelve-foot screw to change size by sixty-five thousandths. This is over a sixteenth of an inch. A twelve-foot screw is typical for a ten-foot axis. Even a ten degree temperature change will increase the length of the screw by over twenty thousandths of an inch. If you are trying to hold a few thousandths of an inch, it is virtually impossible unless you can keep the temperature of the machine absolutely constant. A one-degree temperature change moves the machine more than two-thousandths, but the actual effect is even worse.

A machine tool is made up of castings, weldments and components each built from different types of steel, aluminum and other materials. Some steel, for example, expands at twice the rate of other steel. Also, the materials are interconnected so that between variations in material expansion and interactions between materials in the structure, the machine will twist and distort with changes in temperature. The actual effect of these changes tend to be magnified by the interaction of components to one another. Also, since these interactions are difficult to predict, it is generally not possible to compensate for them using temperature compensation.

As the machine expands with temperature, doesn't the part also expand? Yes it does, but not at the same rate.

Plastic materials have a coefficient of expansion of from 40 millionths of an inch per inch per degree for phenolic to 125 millionths of an inch per inch per degree for polyethylene. The parts may expand up to ten times faster than the machine. Because of the molded shape of the part, it also doesn't expand evenly. Just like the machine, the part also warps and twists as it expands.

With temperature change, the machine grows and twists and the part grows even faster and warps. It is not surprising that the relationship of the program path to the part will also change. The net result is a decrease in accuracy.

Temperature changes can have other effects. As the router bit cuts, it generates a certain amount of heat. This heat first heats the bit and then the heat is transferred to the collet, then the tool holder, then the spindle shaft itself.

As the spindle shaft heats, it becomes longer. Since the spindle housing generally doesn't heat as much as the spindle shaft, it doesn't grow at the same rate. If the spindle bearings at each end of the spindle shaft are rigidly mounted in the housing, the expanding spindle shaft will exert a tremendous load on the bearings and they will fail.

To prevent this, most spindle manufacturers will rigidly mount the lower bearing of the spindle and will allow the upper bearing to float up and down the few thousandths required to keep shaft expansion from loading the bearings. The reason that they rigidly mount the lower and not the upper bearing is to prevent the shaft expansion from changing the location of the tip of the tool too much.

If the spindle shaft is changing size, the location of the tool with respect to the spindle housing and its mounting will also change. Even with the lower bearing rigidly mounted, this size change will affect accuracy.

The distance from the lower spindle bearing to the end of the collet can be four to five inches on many machines. Add to this three or four inches of tool length and you have a total distance of up to nine inches.

The temperature change here is greater than changes in room temperature. A tool can go from sixty degrees before it begins to cut up to a hundred and thirty degrees or so when it is operating. This is a seventy-degree temperature change.

Using a ten millionths of an inch per inch per degree expansion rate, this results is a length change of almost six and a half thousandths of an inch.

At this time you should be starting to get some appreciation for the difficulty in achieving tight accuracy. You should also be wondering how any level of commercial accuracy can be achieved. The final result is not quite as bad as it sounds.

Each of the factors discussed here will affect the ability of the machine to accurately achieve a point in space. Each factor, however, will affect position in a random fashion. Some will move the head in the plus direction while others move it in the minus direction. The final result is that, in general, the accuracy actually achieved is better than it should be considering all the factors affecting it. This random effect on accuracy can also hold a trap if we try and improve accuracy by improving certain factors of the machine.

For example, we may determine that a rail is not perfectly straight. Laser straightening the rail may actually result in a lower level of accuracy if the curved rail compensated for sag in the bridge due to the Z Axis weight. Stiffening the bridge may result in more twist due to temperature change and even more random inaccuracy.

It is quite easy to "chase your tail" in trying to achieve higher levels of accuracy from your machine.

Can you obtain a high level of accuracy on a large machine by controlling all the factors that affect machine accuracy? The answer is yes. I have personally been involved in two projects where this was done successfully. The first of these projects was a defense related program to repair damage to honeycomb wing sections of fighter aircraft. It consisted of two large five-axis machines. Each machine was to achieve a positioning tolerance in free space of plus or minus two and a half thousandths of an inch.

The first machine was to cut out and remove the damaged section of the wing under CAD/CAM control. The second machine was to cut a replacement piece that was bonded into the cut out area from the first machine. The match between the cut out and the replacement plug was critical.

The machines were built in the normal manner, however, the rails were straightened very accurately using a laser interferometer. Then the rails were bowed to compensate for bridge sag so that the path of the head was as accurate as possible.

Machines foundations were over three feet thick and were secured to bedrock and isolated from the main factory floor so that no movement from the floor could be transmitted to the machine foundation and thus to the machine itself. The machines were surrounded by a temperature insulated room in which the air temperature was controlled within one degree.

Overall, the installation was successful and was able to consistently achieve a tolerance of less than two and a half thousandths. The cost of these two machines and their installation was just over $22 million.

The second installation was not one that used our machine. It was a single, custom built machine with a three-foot by three-foot table and needed to achieve better than one-thousandth of an inch tolerance in five axis.

The base of the machine was hollow and filled with circulating oil that was kept at a set temperature within one tenth of a degree Fahrenheit. The spindle motor was also surrounded by a circulating oil jacket and was temperature controlled. The entire machine was enclosed in a room whose air temperature was kept within one-tenth of a degree Fahrenheit. This room was enclosed in a second room that was kept to within one degree.

The machine used dual feedback, the most accurate system being a laser interferometer on each axis. They used a diamond as the cutting tool and a Cray IV Supercomputer as the control system. The entire package cost $55 million and was only trying to hold a thousandth of an inch in five axis.

I have seen companies advertise a half a thousandth accuracy for a CNC router operating in a normal factory environment. The only way they can do this without being laughed out of the industry is that most companies really don't understand accuracy.

Although I can't comment on steps taken by other five-axis machine manufacturers, we devote a large portion of the manufacturing process to trying to achieve the best accuracy performance possible without dramatically increasing the price of the machine.

We start by machining the rail supports on a large machine built to aerospace standards and laser straightened. Using special procedures we make certain that the linear axes are all exactly perpendicular to each other. Once the machine is assembled, we adjust the rails using a laser at the machine head. This compensates for both rail straightness and sag. The lead screws for the linear axes are then measured and calibrated every half an inch to compensate for screw lead error.

On the rotary axes, the worm gear and shaft are machined from a single piece of steel. First the shafts are turned and ground, then using the precision ground shafts, the worm gear is cut. A similar process is used to cut the spur gear. A gear blank is first attached to the turned, ground shaft and then the gear teeth are cut using the precision shaft. This is intended to assure that the gear teeth and shaft are perfectly concentric. Even with these precautions, it is necessary to go through a lapping process where the worm and spur gears are worked together until there are no high spots and backlash is constant through the entire rotation.

After assembly, each of the rotary axes are compensated to remove any slight variation that might still exist. The head is then accurately aligned using our ten point alignment procedure.

We then perform a variation of certain aerospace cut tests that provide a good indication of the final accuracy of the machine.

After all this effort, it is interesting to note that we do not publish an accuracy specification, only a repeatability number. The reason we decline to publish an absolute accuracy specification is that anyone really knowledgeable about machine accuracy would not ask for such a number. They would instead ask about how many of the areas we have just discussed are handled. On the other hand, a person that asks for an accuracy number will likely use this number as a crutch and an excuse to avoid many of the things required to actually achieve an accurate end product.

I can recall one instance where a company purchased a machine to produce wood patterns. The programs were developed on a CAD/CAM system. Initially, the CAD/CAM system was restricted to three axes so the patterns were machined using only three axes. Three-axis machining eliminates many potential problems possible with full five-axis machining. This worked so well that they bought a second machine.

Then they upgraded their CAD/CAM system to full five axis capability. They decided that they could attack every surface from different angles and expected absolutely perfect blends. The result was as expected.

One example was a flat surface that was programmed with both the rotary axes at a forty-five degree angle for half the surface. Then the head was repositioned to the opposite forty-five degree position and the remainder of the surface machined. The company used an extra long tool and the two surfaces had a mismatch of five to ten thousandths. We were delighted that it was that close, the customer was aghast.

The programmed angles combined with the long tool length, multiplied any rotary positioning error. Even slight errors in tool dimension or length affected the part. Repositioning the head added to this the multiplying effect of rail error. The result was that less than a thousandth or so error in the machine multiplied to become a five to ten thousandths mismatch.

We showed the company how to program the part using their CAD/CAM system to minimize or eliminate the problem by keeping the tool close to perpendicular to the surface being cut. There are many techniques that minimize the effects we have been discussing. They felt that the machine should be able to cut accurately using any type of input and that they should not have to program in any special way to achieve accurate parts.

As you can see, these people really didn't understand accuracy. In these cases, providing an accuracy specification is like offering a guarantee that they will achieve perfectly accurate parts regardless of what they do. This is simply not the case.

After having said all this, I will admit that most companies are able to achieve quite accurate trim paths by using a very practical approach.

Although achieving an accurate position in space from theoretical data is quite difficult, repeating the position once it has been achieved is a lot easier. A good tight machine can generally repeat rather accurately. Many companies are able to accurately produce parts on a production basis once an accurate original program has been developed. To obtain an accurate production program, they first develop a program, either at the machine or using a CAD/CAM system. Then, a test part is run and checked. Any areas that are out of tolerance are adjusted at the machine and another part run. This process is continued until an acceptable part has been developed.

Once this type of program has been refined, baring a significant change in temperature, subsequent parts will be fairly accurate reproductions of the original.

If the program is moved to another machine, it may require additional adjustment depending on the operating differences of the two machines. We generally try and build our machines so that a program can be moved from one machine to another with little or no adjustment provided typical commercial tolerances are desired.

Once you have developed an acceptable program you can expect it to run essentially unchanged. I have heard of instances where a company needed to adjust a program several times a day to keep it in tolerance. In this case, most likely there is some type of problem with the machine. A typical good quality five-axis CNC router will be able to consistently produce identical parts once an acceptable original program has been developed.

Although this simple procedure works very well, in today's increasingly theoretical world, more and more companies that don't really understand accuracy are trying to achieve unrealistic performance directly from CAD data. I can only hope that their disappointment does not significantly detract from a tool that can make the entire industry more competitive. Like any tool, much of its capability is not inherent in the tool itself but is in the hands of the person using it.

Chapter 7
Economics

Economics

How much does it cost to trim a part on a CNC router? We are going to try and answer that question by offering you a method of calculating the cost in your factory. Generally the alternative to trimming with a CNC router is to trim the part by hand using a guide fixture or jig. The formula I will offer here will calculate the cost of this method also, so you can compare the result.

We are going to begin with a simple basic formula to define the cost of processing a part through a production center. This formula is deceptively simple when you first look at it. Once we begin to examine it more carefully, we will see that it is actually much more involved than it first appears.

The cost to process a part through a production center is:

Processing Cost = Machine Cost + Labor & Overhead + Tooling Cost + Handling Cost

In this formula we focus on the manufacturing cost through a production center. In the case of trimming a plastic part, material cost will only become significant if it somehow changes through the production process or if it varies when comparing one production process to another. Should this occur, material cost variation can easily become the most significant single cost factor.

There are developments happening today that could significantly tip these cost calculations. Thermwood has developed an effective five-axis chip collection system for its CNC routers. Normally, trim chips have simply been allowed to fall on the machine and floor where they were then swept up and hauled away. Since they can easily become contaminated on the floor, it is generally not worth the risk of trying to reuse this material.

When calculating costs, the labor cost of cleaning up the machine area must also be considered. When trimming thermoplastics, if a five-axis chip collection system is used, the material picked up by the collection system can generally be reused. Thus, the reduced clean up labor as well as the material savings must be worked into the calculations.

Let us now analyze each of these costs in more detail.

Machine Cost

Processing parts at slow speed through an expensive machine will cost more than processing it through a low cost machine at high speed. At least this is the general feeling of most of the industry. In truth, however, this may not necessarily be true.

For continuous production of a single part, these factors are, in fact, the major points to be considered. As the production run size decreases, as is common in processes such as thermoforming, other factors gain in importance and eventually as the batch size gets small, these other factors overshadow the speed and machine cost.

The actual machine cost per part is a combination of factors. A general formula for machine cost per part is:

Machine Cost = Capital Cost + Maintenance Cost + Floor Cost

Capital Cost

Capital cost is the actual cost of owning the machine during the time it is processing the part. But even here, it is not quite that simple. If a machine processes a part in one minute, the cost of processing the part may not be the cost of owning the machine for one minute. If the machine is only used to make one part per year, the machine cost for that part is the cost of owning the machine for one year, not the cost for one minute.

There are several factors that must be considered. The first cost is the actual price of the machine. This is the purchase price plus installation. This is generally the amount that the accountants capitalized on the company's books. If a machine costs $90,000 and it cost $10,000 to install it, they will capitalize $100,000. The machine purchase is not an expense. You paid $100,000 in cash but received a machine worth $100,000. Instead of cash, which is an asset, you have a capital asset.

The accountants will now depreciate the machine. This is based on the theory that each year the machine will be worth less than the previous year. The amount of value that the machine loses each year is called depreciation and this is an expense.

There are several ways of depreciating a piece of machinery. First, a useful life is assigned to the machine. This life is strongly influenced by both tax law and

Generally Accepted Accounting Practices. If a company is making money, they will generally want the life to be as short as possible. This makes the depreciation as high as possible. This higher cost from depreciation reduces earnings and thus reduces taxes. Obviously, the taxman would like as long a life as possible for the opposite reason. If a company is losing money, they might want a longer life so that depreciation cost is reduced helping reach for break even. In actuality, there is not much freedom in assigning a depreciation life to a machine.

So, from the standpoint of actual machine cost, the depreciation should provide a reasonable annual expense level that we can use in our formula. But again, it is a little more complex than that. If we financed the machine, we also have to pay interest on the amount of money we borrowed. This interest is an expense in addition to the depreciation and must be included. If we did not finance the machine, we do not have the interest expense, however, we still have a certain cost for the use of capital. If we did not use our money to buy the machine, we could have put it in the bank and earned a return. The common term for this is the "opportunity cost" of money. Since we did use the money, we will not earn that return and so the money we didn't make is actually a cost for the machine. Another way of looking at this is if we didn't buy the machine we would have this money. Since we did buy it we don't have the money, so it should be treated as part of the cost of owning the machine.

To calculate the opportunity cost, we take the interest rate times the average amount of money tied up in the machine. Using straight-line depreciation, the average amount of money tied up in the machine is half the original cost of the machine. If we finance the machine, we will be making payments and so the actual interest will be highest at the beginning of the payments but will be reduced as the principal is paid off. The average interest is then the interest rate times half the original cost of the machine. As principal payments are being made, the equipment is being depreciated, so for a machine that is financed, we can assume that there is no opportunity cost of money for the principal.

We could be somewhat more accurate by using the exact depreciation level and interest cost each year, however, the results will then depend on where in the life cycle of the machine the calculations are made. Since we are trying to compare basic manufacturing structures to each other, these additional complexities will only make it more difficult to determine the most cost effective methods.

For our purposes we will assume that the interest cost in either case is the interest rate times half the cost of the equipment.

We are now to the point where we can state:

Capital Cost = Depreciation + Interest (paid or lost)

Maintenance Cost

Maintenance cost is relatively simple to calculate. It is the cost of maintaining the machine over a period of time. To get a true cost, you must take the entire maintenance department cost, labor, overhead, parts, etc. and distribute the appropriate part of this entire cost to each of the machines using some formula. Assuming the formula is fair, you will come up with maintenance cost for a period of time, i.e. a year.

Floor Cost

Floor costs is another concept that most managers don't pay much attention to, but it can have a major effect on both costs and attitudes. The formula for floor cost is:

Floor Cost = Cost of Factory Space + Carrying Cost of Inventory

The theory here is that there is a certain cost associated with the space occupied by a production center. To the processing cost we should also add the cost of the space and the carrying cost associated with any material inventory required to support the production center.

The same arguments used for determining the capital cost of the machine can be used to determine the capital cost of the space occupied by the machine and it's supporting inventory. In addition, we must add interest costs of the inventory. Obviously, if the inventory wasn't required by the production center, we wouldn't have to pay to keep it in place and costs would be lower.

As you can see, this is getting quite complex and we aren't even finished with the first factor in the Processing Cost. It is important that all of this be considered. Most of these factors have very little effect when production runs are very long and they have generally been discarded as immaterial.

When production lots become smaller, as they sometimes do in thermoforming, these factors actually become dominant. If you ignore them, you will be totally unable to see major potential for improvement.

At this point, we have examined each of the factors needed to determine Machine Cost. As an example, let's calculate the annual Machine Cost for a sample machine. Although these numbers may vary quite a bit, for our purposes let us use a machine and installation cost of $100,000 and an annual interest rate of 10%.

Straight Line Depreciation $100,000, 10 Years = $10,000 /yr

Interest cost = 10% X $50,000 = $5,000 /yr

Capital Cost = Depreciation + Interest = $15,000 /yr

For purposes of this example let us assume a maintenance cost is 5% of the machine cost:

Maintenance Cost = $5,000 /yr

This number will depend greatly on the machine you selected and the size and efficiency of your maintenance department.

Finally, we will calculate Floor Costs assuming a building cost of $25 per square foot for a building with a 25-year life, straight-line depreciation. Let us assume that at an average, $10,000 worth of inventory sits in front of the machine and that the machine and inventory occupy approximately 2,500 square feet.

Space Cost = 2,500 Sq Ft X $25 /Sq Ft cost = $62,500

Depreciation = Space Cost ÷ 25 Year Life = $2,500/Yr

Capital Cost = Average Space Cost X 10% = $31,250 X 10% = $3,125 /Yr

Inventory Carrying Cost = $10,000 X 10% = $1,000 /Yr

In this case, the Floor Cost is:

Floor Cost = Cost of Factory Space + Carrying Cost of Inventory

Floor Cost = $2,500 + $3,125 + $1,000 = $6,625 /Yr

We can now determine the Machine Cost per Year as follows:

Machine Cost = Capital Cost + Maintenance Cost + Floor Cost

Machine Cost = $15,000+$5,000+$6,625 = $26,625/Yr

From here we must now try and determine how much of this amount to charge against a particular part. We could simply divide this total by the total number of hours in a year to determine the cost per hour. This, however, doesn't work. For example, if we only produced one part per year, that part would only be charged with a tiny fraction of the real cost of owning the machine. To properly analyze the production structure we must charge the cost of owning the machine to the work it actually produces. Obviously, the more parts the machine produces the lower the cost per part and the fewer parts it produces the higher the cost per part. We have formulas here to determine exactly how much these differences are.

There are a couple of ways of approaching this question. First, we can try and estimate the total number of hours per year that the machine is actually producing parts and divide the annual machine cost by this number to obtain a cost per hour. We can then multiply the processing time for a part by this number to obtain the cost of processing a part through this machine. This approach will yield reasonable results provided that the time required to set up the machine for any part is essentially the same and that the production lot size is also generally the same. If, on the other hand, it requires an hour to set up the machine for a two part run, this approach will yield a cost which is much lower than it should be.

A better approach is to determine the total number of hours each year that the machine is either being set-up or running and then dividing the annual machine cost by that total to obtain the machine cost per hour.

In our example, let us assume that the machine is either running or being set-up 75% of the total time. Ignoring holidays and overtime, which tend to cancel each other out, we can say that:

Total Hours per Year = 40 Hr/Wk X 52 Weeks = 2,080 Hr / Yr

If we use the machine 75% of the time:

Machine Hours per Year = 75% X 2,080 = 1,560 Hr/Yr

We can now determine the machine cost per hour as follows:

Machine Cost / Hour = \$26,625 ÷ 1,560Hr = \$17.07/Hr

We still have to determine the machine cost per part. This requires that we determine the time required to set up and run a lot of parts and then divide this total cost by the number of parts processed to determine the cost per part.

Let us continue our example by calculating the machine cost per part for two different sized production runs, the first with 1,000 parts and the second with 100 parts. We will use a 1 hour set-up time and a 10 second per part processing time. First, let's calculate the cost for a 1,000 part run. The total time required for this run is one hour for set-up plus 10,000 seconds or 2.78 hours for run time. The cost per part is as follows:

Cost of Run = (1 Hr Set-up + 2.78 Hr Run Time) X
\$17.07 /Hr = \$64.52

Machine Cost per part = \$64.52 ÷ 1,000 parts =
\$0.065 per part

Now, let us perform the same calculation for a 100 part run. In this case there is still a 1 hour set-up time but run time is only 1,000 seconds (100 parts times 10 seconds per part). This is .28 hours. The cost per part is then:

Cost of Run = (1 Hr Set-up + .28 Hr Run Time) X
\$17.07 /Hr = \$21.85

Machine Cost per part = \$21.85 ÷ 100 parts =
\$ .22 per part

In these examples, a part run in quantity 100 is THREE TIMES more expensive than a part run in 1,000 quantity. If the machine were ten times faster, which would be seen by the industry as a major advantage, per part cost would still be $.18. In this example the conventional wisdom that a faster machine is always much better absolutely doesn't work. At this point we have only examined the machine cost, one of four cost factors in the overall processing cost of the part. Once we have examined the rest of the cost factors we will perform more of this type of analysis to try and understand how different approaches can affect the processing cost. This example is simply a quick view to show the magnitude of differences that result from changes in volume and to point up that accepted truths in the industry may not hold up to careful analysis.

Labor and Overhead

The next major factor in determining processing cost is labor and factory overhead associated with that labor. The labor cost normally associated with a production process is called "Direct Labor" because it is a direct cost of production. Factory overhead is the general factory expenses that cannot be associated with a specific production run. This includes light, heat, electricity and certain supplies. The most common way of accounting for factory overhead is to take the total factory overhead for a period of time and divide that by the average number of direct labor hours worked during that period. The result is an overhead rate per direct labor hour. Each time a direct labor hour is used to produce a product, the overhead cost associated with that direct labor hour is also charged to that product.

Some factories pay their people based on the number of parts that they produce. They are paid for each piece produced. In this case, it is generally not possible to associate overhead with a direct labor hour. This situation can be handled by dividing the factory overhead by the total production payroll and determining an overhead rate per dollar of direct labor payroll.

The true per hour cost for an employee is actually higher than the per hour rate that the employee is paid. To the direct pay, you must add the cost of fringe benefits such as insurance, holidays, vacation and the like. Add to that the factory overhead rate per hour and this total cost per hour is commonly called the burdened labor rate and includes the direct pay, cost of fringe benefits and overhead rate. This number is generally developed by the accounting department of most companies and will generally be adjusted once or twice a year. For our purposes we can simply use the accepted number in your factory, although there are pricing techniques which rely on unbundling and treating each of the components separately.

To determine the labor and overhead cost for processing a part through a production center we must consider three separate cost areas. These are set-up cost, processing labor cost and material handling and removal labor cost. For this analysis, we will consider the set-up labor and overhead cost and the processing labor and overhead cost in this area and we will address the material handling labor and overhead costs in the section on handling costs.

The formulas for this are straightforward:

Labor and Overhead Cost = Set-up Labor and Overhead + Processing Labor and Overhead

If the same people that set-up the machine also run the machine, it is simply a matter of multiplying the total man hours devoted to setting up and running the production run by the burdened labor rate. In many factories, however, the people that set-up the machine are not the same people that run the machine. In many cases, the pay rates are different and so this difference must be considered. For this reason, we have separated the two activities, set-up and processing, into two separate areas.

Some companies consider setup people indirect labor and do not associate factory overhead with their time. Although this is a valid approach from an accounting perspective, it will tend to mask the true costs associated with short production runs. We will assume that setup labor is direct labor in this analysis, however, if you feel it necessary to treat it as indirect labor, a separate indirect labor rate can be used with the software package covered later in this book.

When two different groups are required, another factor may occur. If the machine operator simply watches while the machine is set-up, his or her cost must be added to the cost of setup. If not, then a certain amount of time is lost returning to the machine or moving to another assignment. On short production runs, these factors can become significant. In presenting examples, I will mostly ignore these effects, however, they will increase, perhaps significantly, the cost of shorter production runs. The differences should be dramatic enough without these additional factors that we can make our points without the extra complexity of these additional considerations.

Let us use these formulas to calculate labor and overhead costs for our example. In the example we will use a burdened labor rate of $30 per hour. Many companies will have a rate well under this number and some may have a higher rate. As we develop some of the new production methods, they will use fewer people than current systems. Fewer people mean fewer direct labor hours to absorb existing overhead costs. This translates into higher overhead rates per hour and thus higher burdened labor rates. By using a $30 number, we are using a reasonable average and can ignore this effect. In actual practice, however, as we become more productive and require fewer direct labor hours, those hours will cost more because the overhead will be spread over fewer total hours.

We will use the same rate for both setup and processing, although in many cases the setup rates are higher. This will tend to make short production runs look slightly better than they really are, however, they will look so bad anyway that the slight difference will not be a factor.

Let us again look at both a 1,000 part run and a 100 part run. For a 1,000 part run, it requires an hour set-up and 2.78 hours processing time so the cost is:

Labor and Overhead = (1 Hr Set-up + 2.78 Hr Run Time) X $30 /Hr = $113.40

Cost per part = $113.40 ÷ 1,000 parts = $0.11

Now let's look at the same situation with a 100 part run:

Labor and Overhead = (1 Hr Set-up + .28 Hr Run Time) X $30 /Hr = $38.40

Cost per part = $38.40 ÷ 100 parts = $0.38

Again, the cost per part for the short run is over three times higher even though the part was run on the same machine at the same speed.

Tooling Cost

Most processes use some type of tooling. There is a cost associated with specifying, purchasing, mounting, sharpening and replacing these tools. We have included tooling in the base formula because in some circumstances the cost of tooling differs enough between alternate production structures that it must be considered. For our purposes, we will be comparing different factory structures using essentially the same machining processes. In the processes we will examine, there will be little difference in tooling costs from one structure to another. Therefore, we will not take tooling costs into consideration as we examine the various alternatives.

Tooling not only consists of router bits, but also includes trim or guide fixtures, if the process requires these. It is common to spend a substantial amount of labor adjusting and repairing these fixtures and jigs and the cost of this effort must be considered.

If you are trying to determine the specific cost of a part, you will need to carefully examine the tooling costs. In many processes, the tooling cost is a significant part of the overall processing costs and must be examined carefully. Also, if you are going to use these formulas to evaluate a process or production structure where tooling cost, life or other factors are not essentially the same, you will need to include a tooling analysis as part of your evaluation. When you do these analyses, examine not only the cost of the tools but also the labor and machine time required for the tooling change, assuming this change occurs during a production run. If tool changes occur only during set-up, the tool change time should be included in the overall set-up time.

Handling Cost

This is an important concept based on the traditional layout of a manufacturing plant. There is a cost associated with moving the parts from one production center to the next. For purposes of this analysis, we will add to each production center the costs associated with bringing the material to the production machine from wherever they were located before. The cost of removing the finished parts will be associated with the next production center.

To determine the cost associated with handling the product we will use the formula:

Handling Cost = Equipment Cost + Labor and Overhead

For purposes of these calculations, the equipment costs include the depreciation and capital costs as described earlier, however, now these are applied to the material handling equipment. This can be a fork truck, conveyor or other machines.

It is important to take this cost into consideration since a structure that does not require material handling, for example, a production cell, does not incur these costs between production centers.

The best way to determine the handling cost per part is to determine the total handling cost required to move an entire production batch and then divide this number by the number of parts that were moved. Here again, batch size will have a major impact on cost. For example, if you normally move parts using a pallet and fork truck, and let us assume that you can stack 500 parts on a pallet, then the cost to move one part or five hundred parts is the same. That is the cost to move one pallet. On a per part basis, however, the cost per part is five hundred times more if you are only moving one part.

Let us plug some numbers into the formulas to continue with our example. Let us assume that the fork truck or material handling equipment has a cost of $20,000. In this case we will use machine cost and capital cost but we will not assign specific floor space to the material handling effort.

Machine Cost = Capital Cost + Maintenance

Capital Cost = $2,000 depreciation + $1,000 opportunity cost + $1,000 maintenance = $4,000/Yr

Capital Cost per Hour = $4,000 / 1,560 = $2.56/Hr

We will again use the $30 burdened labor rate to calculate labor and overhead costs. The total cost per hour for equipment, labor and overhead is $2.56 plus $30 or $32.56. Let us assume that it takes 10 minutes (.17 Hr) to move a pallet and we can put 500 parts per pallet, maximum.

The calculations for handling 1,000 parts are:

Handling Cost = 2 trips X (.17 Hr X $32.56) = $11.07

Handling Cost per Part = $11.07 ÷ 1,000 = $.011

Now let's do the same calculation for 100 parts. This time it only takes one trip.

Handling Cost = 1 trip X (.17 Hr X $32.56) = $5.54

Handling Cost per Part = $5.54 ÷ 100 = $.055

The impact here is even greater. It costs five times more per part to move the 100 part lot as it does to move the 1,000 part lot.

Let us take a look at the total processing cost. Here is our basic formula:

Processing Cost = Machine Cost + Labor & Overhead + Tooling Cost + Handling Cost

We will ignore the tooling cost so plugging in the numbers we have developed yields the following processing costs:

1,000 Part Run

Processing Cost/Part = $ .065 + $ .11 +$ .011 = $.186

100 Part Run

Processing Cost/Part = $ .22 + $ .38 + $ .055 = $ .655

Eighteen cents versus sixty-five cents for the same part run at the same speed through the same machine. The difference here is structure. In this simple example using reasonable numbers, the cost of one structure is over three times more expensive than the other. At 1,000 parts or more, the initial structure actually makes some sense. If your volume has dropped, however, the old structure no longer works. If you are going to produce parts in 100 quantity you will need to find a structure that will at least restore the costs associated with your old structure. If you are just starting out and these are the levels you can operate at, you still need to compete with companies that operate at higher levels. No one can operate a business where costs are three times higher and survive very long.

To understand this cost relationship a little better we have plotted the cost per part versus the lot size using the numbers from our example. Here is the result of that plot:

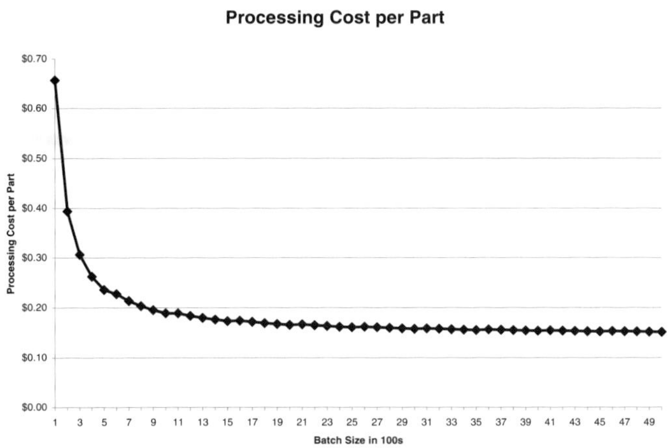

You will notice that as the volume drops to 100, the processing cost per part skyrockets. As the lot size drops below 100, costs rise even further until a single part run costs almost $68 to process. As you can see, the cost seems to settle out about 20¢ after the lot size increases above 2,500. You will also notice that once the lot size drops below about 500 parts, cost per part rises substantially faster.

Changing the costs of the various parameters will affect the way this chart looks. The chart presented above is simply designed to illustrate the impact of lot size on cost in this particular example. Each set of parameters will have a lot size where the cost per part rises substantially.

This analysis does not take into account another factor which has little effect at large lot sizes but can have a very large cost effect at very low lot sizes. This effect is the tendency to scrap a part or two during set-up on most machines. Scrapping a couple test parts during set-up will have a minimal cost effect on a thousand-part run. On a hundred-part run, this amounts to two percent of the batch. If the part must run through four or five production centers, this two-percent becomes eight to ten percent which is becoming significant. On even smaller batch size this can become an extremely large cost. The cost is also higher if more than a couple of parts are scrapped at each set-up.

Another cost effect is associated with the practice of running a few extra parts with each batch just in case a few extra parts get scrapped in the set-up or assembly process. Obviously it is more cost effective to run a few extra parts than to set-up each of the machines to make up for a possible shortfall. Again, an extra ten or fifteen parts has little cost effect on a large production run. These extra parts however are ten to fifteen percent of a hundred-part run and are an even higher percent of a smaller batch. These extra parts are either scrapped at the end of the run or must be labeled and stored for the next time the part runs. Either approach adds cost.

The formulas and underlying ideas presented in this chapter can be applied to any production machine. They are not restricted to CNC routers. They should provide you with a means of calculating processing cost for each type of machine in your factory.

To make these calculations easier, we have supplied, with this book, a simple software program that will operate on a PC running Windows 95. Using this program, your actual numbers can be input into the formulas and you can see the actual costs of processing your product. It is easy to vary each of the parameters and see the effect on the processing cost.

If you sit down with this package and some typical parts for an hour or two, you will develop a very good feeling for how cost vary as different parameters vary within your plant. This should develop an intuitive feeling that is accurate and this intuition should provide you with guidance in your day to day management decisions.

In the remainder of this chapter we will detail the use of the software program.

The program is intended to operate on an IBM PC or compatible running Windows 95. To install the program, place the enclosed floppy disk in the appropriate drive and type:

A:/setup

Where "A" is the drive number. The system will then guide you through the steps necessary to complete the installation.

Once installed and started, the program will display the screen shown here.

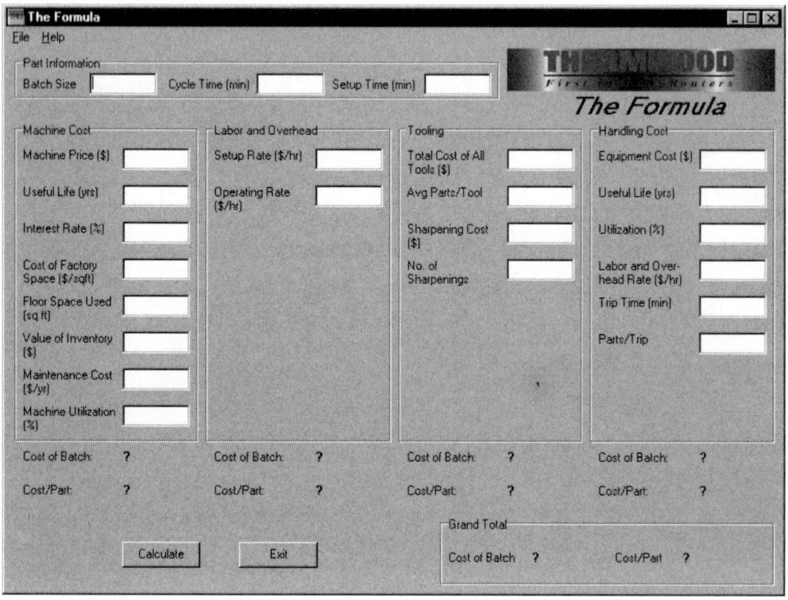

This screen displays the formula as described in this chapter. At the top, three fields define the batch to be run. Each of the four cost areas, machine cost, labor and overhead, tooling cost and handling cost have a separate area where all the factors can be input. As the necessary factors affecting the cost area are input, the batch cost and total cost for the area will be displayed at the bottom of the area.

Each cost area has several fields or input boxes. Each field contains a single number that affects the manufacturing cost. You will notice that when you move the mouse pointer over one of these fields, an information box appears detailing the information that should be entered in the box. This is intended to give you a quick review of the appropriate input number. These information boxes are not an adequate substitute for the more detailed information included in Chapter 1 of this book. Make certain that you have read and that you understand the concepts in Chapter 1 before you attempt to use the software package to develop cost data.

At the top of the screen you will find the Part Information area. This area contains three input fields.

Batch Size – This is the total number of parts that will be run using the current set-up.

Cycle Time – This is the time in minutes required to load, process and unload a part.

Setup Time – This is the time in minutes required to set-up the machine for the production run.

Machine Cost

The first cost area is machine cost. There are eight input boxes required to determine the machine portion of the processing cost.

Machine Price – This is the cost of the machine in dollars that has been capitalized and is being depreciated.

Useful Life – This is the period of time in years over which the machine is being depreciated.

Interest Rate – This is the rate of interest that you could have earned on the money used to buy the or the interest you would pay to borrow money to purchase the machine.

Cost of Factory Space – This is the cost to build or purchase the factory floor space per square foot.

Floor Space Used – This is the amount of factory floor space required to house both the machine and the inventory required to support the machine.

Value of Inventory – This is the value of the average inventory associated with the machine in dollars.

Maintenance Cost – This is the average material, labor and overhead cost associated with maintenance of the machine for one year in dollars.

Machine Utilization – This is the percent of the total time based on a single shift, eight hour a day operation (2,080 hours/year) that the machine is either being set-up or is running production. A two-shift operation would be 200%, a three-shift operation up to 300%.

Labor and Overhead

The next area is labor and overhead. In this area you will input the hourly direct labor and the associated factory overhead. Review the description of the labor and overhead rate presented earlier in this chapter to make certain that you use the full burdened labor rate.

Setup Rate – This is the hourly direct labor cost and the associated factory overhead. If the machine operator must remain idle while the machine is set-up by another, the rate is the total for both the set-up labor and the operator labor.

Operating Rate – This is the per-hour direct labor cost and the associated factory overhead for the machine operator. If more than one operator is required, the number is the total per-hour labor and overhead for all operators required to run production on the machine.

Tooling

The next area is tooling. In this area we will try and determine the tooling cost associated with processing a part. If a part requires more than one tool, we will average the cost and life information to try and develop a reasonable approximation of the tooling cost.

Total Cost of All Tools – This is the total cost to purchase all tools required to process the part.

Avg Parts/Tool – This is the average number of parts that can be processed by a tool.

Sharpening Cost – This is the average cost of sharpening a tool. If you are considering a hand trim fixture, this can also be the cost of maintaining the fixture on a regular basis.

No. of Sharpenings – This is the number of times a tool can be sharpened before it must be discarded.

Material Handling

The final area concerns the cost to move material from the previous production center to the current production center. There are six input fields in this area.

Equipment Cost – This is the cost of the fork truck or other material handling equipment required to move material to the production center.

Useful Life – This is the period of time in years over which the material handling equipment is depreciated.

Utilization – This is the percent of a single shift during which the equipment is being operated.

Labor and Overhead Rate – This is the per-hour direct labor cost and the factory overhead associated with the material handling effort.

Trip Time – This is the average time in minutes required to move a skid of material from the previous production center to the current production center and then return.

Parts/trip – This is the maximum number of parts that can be moved in a single material handling trip.

After all input fields have been entered, press the "Calculate" button and the costs will be re-calculated using the current numbers.

Now that you have the basic tool, it is time to play with the numbers. Set up the package with a set of typical numbers for your company. Now, play with the numbers. Change a parameter, batch size for example and see how the cost changes. Change another parameter, machine cost of setup time and again see the part cost change.

Within an hour or two, you should begin to develop a new intuition about how various factors and combinations of factors affect your costs. Playing with the different cost parameters you should be able to develop a real understanding of what different decisions about your operation of your factory will do to your costs.

To develop an even more detailed understanding, perform the calculations for a series of different machines and see how combining operations will affect costs.

This exercise will take some time and effort but the final result will be well worth the effort. The understanding you develop will provide some guiding principals that will make the transition to a more profitable company easy to understand. The intuition will make the difficult decisions clear.

INDEX

INDEX

F

G

H

I

L

M

N

O

P

R

S

T

U